Nelson Advanced Modular Science

Applied Plant and Animal Biology

JOHN ADDS•ERICA LARKCOM•RUTH MILLER

Thomas Nelson & Sons Ltd
Nelson House
Mayfield Road
Walton-on-Thames
Surrey KT12 5PL
United Kingdom

First Published by Thomas Nelson & Sons Ltd 1999
ISBN   0 17 448270 1
9 8 7 6 5 4 3 2
01 00 99

Page design and artwork by Hardlines, Charlbury, Oxfordshire

Printed in Croatia by Zrinski Printing and Publishing House, Cakovec

Picture Research by Image Select International Ltd

Publishing team
Acquisitions: Sonia Clark
Staff editorial: Simon Bell
Freelance editorial: Liz Jones
Production: Suzanne Howarth

Acknowledgements

The authors and publishers would like to thank the following for
permission to reproduce the following copyright images.

Artwork
Blackwell Science: figure 4.3 and table 4.4 from Croston, D. and Pollott,
G. *Planned Sheep Production*, 1994; figures 4.10 and 4.17 from Peters,
A. R. and Ball, P. J. *Reproduction in Cattle*, 1995; tables 4.6 and 4.7,
figures 4.22, 4.23 and 4.29 from Halley, R. J. and Soffe, R. J., *The
Agricultural Notebook*, 1992; Hoechst plc: figures 4.9 and 4.12; Otley
College: figures 4.11, 4.14, 4.24 and 4.26; D.G. Mackean: figure 4.21,
from Mackean, D.G., *Life Study: A Textbook of Biology*, John Murray
(Publishers), 1981; Longman Educational: figure 4.30 from Speedy, A.
Sheep Production, 1980; Ryder, M.L.: figure 4.31 from 'Wool: the ideal
textile fibre' in *Biologist* (1994) 41 (5); CAB International Publishing:
figures 5.8, 5.9, 5.12, 5.19, 5.20 and table 5.2 from Rose, S.P.,
Principles of Poultry Science, 1997.

Photographs
Erica Clark: figures 4.1a, 4.1b, 4.2a, 4.2b, 4.20, 4.25a, 4.25b, 5.1, 5.2,
5.14, 6.1, 6.7, 6.9, 6.11, 6.13, 6.17; Chris Fairclough Colour Library:
figures 2.6b, 4.28, P6; George Hide/Sparsholt College: figures 6.8a,
6.8b, 6.16; Holt Studios International: figures 1.9a, 2.3, 2.11, (all Nigel
Cattlin), 5.17 (Inga Spence); Image Select/Ann Ronan: figure 4.2a;
Frank Lane Picture Agency: figures 2.7 (S. McCutcheon), 5.16 (Derek
Middleton), 6.15 (W. Broadhurst); Professor John Lewis/Electron
Microscopy Unit, Royal Holloway College London: figures 6.14a,
6.14b; Ruth Miller: figure 7.3b; Science Photo Library: figures 1.6a
(Andrew Syred), 1.6c (Bruce Iverson), 2.6a (Ed Young/Agstock), 3.7
(Rosenfeld Images Ltd.), 6.18 (Simon Fraser); Telegraph Colour
Library: figure 4.32.

The authors are grateful to Richard Bampton of Hadlow College,
Jonathan West of Otley College and George Hide of Sparsholt College
for their help in the preparation of this book.

The examination questions and mark schemes in this book appear by
permission of Edexcel (London Examinations).

Contents

Introduction

As modularisation of syllabuses gains momentum, there is a corresponding demand for a modular format in supporting texts. The Nelson Advanced Modular Science series has been written by Chief and Principal Examiners and those involved directly with the A level examinations. The books are based on the London Examinations (Edexcel) AS and A level modular syllabuses in Biology and Human Biology, Chemistry and Physics. Each module text offers complete and self-contained coverage of all the topics in the module. The texts also include examples outside the prescribed syllabus to broaden your understanding and help to illustrate the principle which is being presented. There are practical investigations and regular review questions to stimulate your thinking while you read about and study the topic. Finally, there are typical examination questions with mark schemes so that you can test yourself and help you to understand how to approach the examination.

In the Option modules of the London syllabuses, we explore applications of Biology, delving into some areas where we really make use of biology in our every day lives. *Applied Plant and Animal Biology* covers two modules, but with essentially common themes. It looks at the way humans have, from the earliest forms of agriculture, exploited and manipulated crop plants and domesticated animals for the production of food. Modern practices in horticulture and agriculture have become increasingly mechanised with concurrent application of scientific principles to maximise productivity. Artificial control over the environment and manipulation of reproduction allows production to be adjusted, in terms of both quality and timing, to meet the fluctuating demands of the mass consumer market.

In the plant chapters, we look at factors affecting plant growth and see how close control of these factors can be achieved in a glasshouse environment. We consider how crop growth is affected by other organisms, through competition from weeds and from disease and insect pests, then look further at measures used for their control. Reproduction in plants is seen firstly as a way of conserving desired varieties, through asexual methods of propagation (including micropropagation), and secondly as a means of changing and improving the characteristics through traditional methods of selective breeding. We take this a stage further and look at the potential for introducing specific changes by gene technology. The animal chapters focus on cattle and sheep, on chickens and on fish raised in fish farms. In each case, consideration is given to aspects of nutrition in relation to productivity, in terms of both quantity and quality of meat, milk, wool and eggs as appropriate. Application of modern reproductive technology gives considerable scope for genetic improvement of the stock and for manipulating the life cycles to meet consumer demands. This is reviewed in the context of traditional breeding programmes. The authors hope that through your study of topics in this book, relating to plants or to animals, you will gain a better knowledge of the biology of food production and develop an understanding of the issues that arise from intensive methods of horticulture and agriculture, especially those related to the environmental impact of human activities and long term sustainability.

The authors

Erica Larkcom B.A., M.A., C.Biol., M.I.Biol., former Subject Officer for A level Biology, formerly Head of Biology, Great Cornard Upper School, Suffolk

John Adds B.A., C.Biol., M.I.Biol., Dip. Ed., Chief Examiner for A level Biology, Head of Biology, Abbey Tutorial College, London

Ruth Miller B.Sc., C.Biol., M.I.Biol., Chief Examiner for AS and A level Biology, former Head of Biology, Sir William Perkins's School, Chertsey, Surrey

Note to teachers on safety

When practical instructions have been given we have attempted to indicate hazardous substances and operations by using standard symbols and appropriate precautions. Nevertheless you should be aware of your obligations under the Health and Safety at Work Act, Control of Substances Hazardous to Health (COSHH) Regulations and the Management of Health and Safety at Work Regulations. in this respect you should follow the requirements of your employers at all times.

In carrying out practical work, students should be encouraged to carry out their own risk assessments, i.e. they should identify hazards and suitable ways of reducing the risks from them. However they must be checked by the teacher/lecturer. Students must also know what to do in an emergency, such as a fire.

The teachers/lecturers should be familiar and up to date with current advice from professional bodies.

Plant growth for crop production

Green plants are **autotrophic** organisms, capable of building up complex organic molecules from inorganic raw materials, using energy from light. This process is known as **photosynthesis**. In addition to light, plants require a supply of carbon dioxide, water and mineral ions. In the presence of the green pigment **chlorophyll**, found in the chloroplasts in leaves and green stems, the inorganic raw materials are converted into carbohydrates, lipids and proteins. These compounds supply the plants with a source of energy for metabolic activities and with the materials from which new cells are made, resulting in growth.

$$6CO_2 + 12H_2O \xrightarrow[\text{chlorophyll}]{\text{light}} C_6H_{12}O_6 + 6O_2 + 6H_2O$$

$$\text{carbon dioxide} + \text{water} \longrightarrow \text{glucose} + \text{oxygen} + \text{water}$$
$$\text{(carbohydrate)}$$

Figure 1.1 Summary equations for the process of photosynthesis in words and formulae

The rate at which energy is stored by plants in the form of organic compounds, which can be used as food, is termed the primary productivity. In the first instance, **primary productivity** depends on the quantity of solar radiation intercepted by the plants, which depends on the latitude, aspect and light quality. In the UK, about 1 per cent of the total solar radiation is intercepted by plants, amounting to about 1×10^6 kJ m^{-2} yr^{-1}. Of this, only between 1 per cent and 5 per cent is absorbed by the chlorophyll molecules, 99 per cent to 95 per cent being immediately reflected, radiated or lost as heat of evaporation. The rate at which chemical energy is stored is termed the **gross primary productivity (GPP)**. Not all of this energy is available to the next trophic level because significant quantities are lost due to respiration and other factors, leaving the **net primary productivity (NPP)**, defined as the increase in dry mass of a crop over a period of time. Productivity is calculated on a yearly basis, thus taking into account variations due to seasonal changes. In the UK, we would expect GPP and hence NPP, to be greater during the spring and summer months than during the winter. As solar radiation energy differs in different parts of the world, so GPP and NPP will vary and it is worthy of note that the primary productivity potential of tropical regions is almost double that of the UK.

Factors affecting plant growth

The rate at which a crop accumulates dry matter, or **biomass**, is dependent on the rate at which photosynthesis occurs. This, in turn, depends on both internal and external factors. The internal factors relate to the leaf anatomy, the amount of chlorophyll present and the activity of the relevant enzymes. The biochemical details of the light-dependent and light-independent reactions of photosynthesis are fully described in Chapter 1 of *The Organism and the Environment*. External factors include the availability and uptake of the raw materials, carbon dioxide, water and mineral ions, together with temperature and light.

Why is primary productivity likely to be greater in tropical regions thani in the UK?
What might limit primary productivity in tropical regions?

PLANT GROWTH FOR CROP PRODUCTION

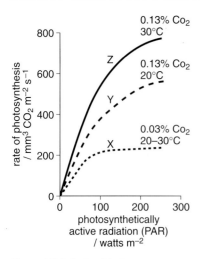

Figure 1.2 Relationship between carbon dioxide concentration, light and temperature

The rate of plant growth at any one time can be affected by any, or all, of these external factors. The rate is always determined by that factor which is in the shortest supply, known as the **limiting factor**. The general relationship between light intensity, temperature and carbon dioxide concentration is summarised in the graph in Figure 1.2. Curve X shows the photosynthetic rate at the normal atmospheric concentration of carbon dioxide. Curve Y shows that by increasing the carbon dioxide concentration, the rate of photosynthesis is increased, indicating that under normal circumstances carbon dioxide concentration is the limiting factor. In Curve Y, the rate begins to decrease as the temperature becomes limiting. Curve Z shows that an increase in temperature has the effect of increasing the rate still further. The graph shows that the environmental factors interact with one another. Limiting factors may also vary with time: very early in the morning, light and temperature may be limiting. It is relevant to note here that control of these conditions can only be achieved when crops are grown in glasshouses.

In the UK, carbon dioxide is the commonest limiting factor in photosynthesis. Light saturation is usually easily reached at below the highest light intensity on most summer days. However, under such circumstances, where light intensity and temperature are both high, **photorespiration** occurs reducing photosynthetic yield by between 30 and 40%. Photorespiration, defined as the light-dependent uptake of oxygen and output of carbon dioxide, occurs because the carbon dioxide fixing enzyme (RuBP carboxylase) also accepts oxygen as a substrate and there is competition between oxygen and carbon dioxide for the active site. Some tropical plants, such as maize and sugar cane, known as C_4 plants, do not suffer such losses. They use an alternative pathway, the Hatch–Slack pathway, for the initial fixation of carbon dioxide and little or no photorespiration occurs, resulting in higher rates of photosynthesis at higher temperatures and light intensities. For more background information on C_3 and C_4 plants and the metabolic pathways of photosynthesis, reference should be made to Chapter 1 in *The Organism and the Environment*.

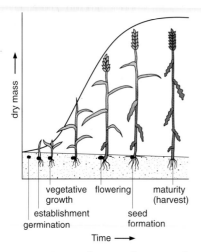

Figure 1.3 Growth curve of an annual crop, e.g. cereal

As photosynthesis occurs, the formation of new compounds within the cells results in an increase in their mass. The cells increase in size due to this increase in mass and also due to the accumulation of water. In addition, new cells are formed, resulting in an increase in cell numbers. So growth of the plant may show as an increase in mass or an increase in dimensions such as leaf area, leaf or stem thickness and leaf density. If measurements of a plant's mass, volume or height are made at regular intervals and plotted against time, a **growth curve** is produced. This curve has a characteristic **sigmoid**, or S-shape, and several distinct phases can be seen (see Figure 1.3).

Measurements of **true growth** are based on irreversible increases in dry mass, cell size or cell number, so a more accurate determination of plant growth would be to use **absolute growth rate**. This is defined as the mean increase in plant biomass, measured as dry mass per unit time. Samples of plants are taken from a population at timed intervals and dried to constant mass. Each sample should contain the same number of plants and a mean dry mass per plant can be calculated. The difference in mass at each time interval is then used to calculate the increase in dry mass per unit time, as shown below.

X_1 g = mean dry mass of first sample taken at t_1 days

X_2 g = mean dry mass of second sample taken at t_2 days

$$\text{Growth rate (G)} = \frac{X_2 - X_1}{t_2 - t_1} \ \text{g day}^{-1}$$

An absolute growth rate curve can be drawn, plotting changes in growth rate against time. This curve is usually bell-shaped and shows the period when growth is most rapid.

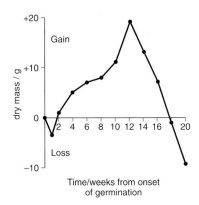

Figure 1.4 Absolute growth rate curve

This method is of limited value as it does not indicate how fast the plants are growing and does not allow comparisons to be made between plants of different sizes or between different crops. Where such comparisons are required, **relative growth rate**, that is the increase in mass per unit mass per unit time, may be used. In addition to enabling comparisons between plants of different sizes, this measurement allows comparisons to be made between different treatments and different environmental conditions.

Usually only certain parts of the plant, such as grains from a cereal, stem tubers of potatoes or leaves of lettuce, are of use and economic value to the crop producer, rather than the total biomass. The quantity of dry matter capable of being harvested is known as the **harvestable dry matter**. In hay or silage for feeding cattle, all the plant biomass which is harvested is used in the product (the economic yield), but in the case of a cereal crop only the grains are used, the rest of the plant having low commercial value. Thus the economic yield is only a proportion of the biological yield, or total biomass. This proportion is termed the **harvest index** and can be calculated by dividing the economic yield by the total biomass.

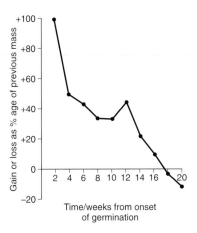

Figure 1.5 Relative growth rate curve

$$\text{harvest index} = \frac{\text{economic yield}}{\text{total biomass}}$$

Quantity, in terms of high yields, may be important, but so is quality, especially if the crop is being produced for a specific purpose. Wheat grown for bread-making or for animal feed should have adequate amounts of protein (**harvestable protein**) and **digestible energy** (that proportion of the energy content which can be digested or absorbed). It is therefore important that the quality of a crop is monitored so that the product reaches the required standard.

When we refer to crop production, we tend to think primarily of the growth of plants for food. Photosynthesis results in the accumulation of stored carbohydrates, such as starch, in the tissues of the plant. The starch stored in tubers and in the endosperm of cereals or the cotyledons of seeds, such as peas and beans, provides humans with an important source of energy. In a stem tuber, such as the **potato**, the thin-walled **parenchyma tissue** is packed with starch grains. Potatoes are an important source of carbohydrate, and hence energy, in many diets.

Temperature
Temperature affects the growth rate of plants by controlling the rate of their

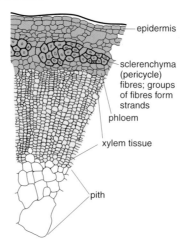
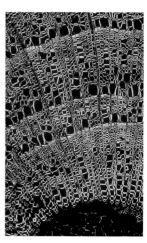

Figure 1.6 (a) Cells of potato with starch grains; (b) Section of flax stem showing fibres; (c) Section of xylem (wood) tissue

Flax, used in the production of textiles, and timber, used for building and furniture manufacture, are both plant products derived from lignified tissues. Flax is the term given to the bundles of sclerenchyma fibres found in the cortex of the stem of the flax plant, *Linum usitatissimum*. These groups of fibres form compact strands towards the outer side of the vascular bundles. Flax is sown in April or May and harvested in August or September, preferably before the seed sets. After harvest, the fibres have to be loosened from the stem, a process known as `retting'. Then the stems are rinsed, dried and crushed, leaving the pure flax in the form of strands which can be combed into fibres. The longest fibres are spun into linen and the shorter ones used in other fabrics and in paper.

The xylem, or woody tissue, of many trees is used for timber. This woody tissue is produced as a result of the activities of the vascular cambium and consists of vessels, tracheids, fibres and xylem parenchyma. The walls of the vessels, tracheids and fibres become lignified and the cell contents disintegrate. The deposits of lignin confer strength and resistance to compression, which makes this tissue of value in the building industry. Many trees are used to produce timber, each having its own characteristic appearance, properties and uses. The timber from the elm, *Ulmus campestris*, has red heartwood and is widely used for making furniture and veneers due to its ability to be bent easily. English elms were affected by Dutch Elm disease in the 1970s, so few trees planted since then have reached maturity.

metabolic reactions. These reactions are enzyme-controlled: generally, an increase in temperature causes an increase in the rate of these reactions, particularly those associated with the light-independent stages of photosynthesis, and hence an increase in the growth rate. Maximum growth at any stage occurs at the optimum temperature for that stage. For the successful growth of crops in any geographical location, a certain number of days at the optimum temperature is required. In the UK, the minimum (base), optimum and maximum temperatures for the growth of most crops fall between 0 ° and 35 °C. The base temperatures for native, or well-adapted, crops are around 0 ° to 1 °C, but for crops such as maize and sugar beet, which originated in warmer climates, the base temperature is around 10 °C. This can be illustrated by reference to winter wheat and rye grass, which are able to grow in March and April, whereas rapid growth in maize and sugar beet does not occur until June or July.

Both soil and air temperatures can affect crop production and they can both vary with the geographical location, height above sea level, season of the year and time of day. The soil temperature depends on the amount of solar radiation which reaches the soil surface. This can be affected by the plant cover and the nature of the soil. Where the soil surface is covered with vegetation, most of the solar radiation is intercepted by the plants and little penetrates. Light-coloured soils tend to be colder than dark soils, which are able to absorb heat more readily. Waterlogged soils, with poor drainage, tend to be cooler than well-drained soils.

Some crop plants, such as sugar beet, winter cereals and perennial rye grass, require a period of low temperatures in the winter before flowering can occur in the following spring (vernalisation). In the UK, normal winter temperatures fulfil this requirement, but if winter cereals are planted late, then temperatures may not be low enough and subsequent flowering and yield are reduced.

Carbon dioxide

Carbon dioxide is an essential raw material for photosynthesis. It combines with the five carbon acceptor molecule ribulose bisphosphate (RuBP) in the light-independent stage of the process, resulting in the formation of carbohydrates. Carbon dioxide enters the aerial parts of the plant, the leaves and green stems, along a diffusion gradient via the stomata. During the hours of daylight, there is a higher concentration of carbon dioxide in the atmosphere than there is in the mesophyll tissues of the leaves, so carbon dioxide enters by diffusion. Oxygen is produced as a waste product of photosynthesis and diffuses out of the leaf. The carbon dioxide concentration in the atmosphere is around 0.03 per cent, so on bright, sunny, warm days, it is the limiting factor in photosynthesis.

Light

Light affects plant growth in a number of ways and so it has a profound effect on crop production, from germination through to flowering and seed production. Many of these aspects are covered in core topics, but it is useful to summarise them here:

- Light is required as a trigger for the germination of the seeds of some crops, especially lettuces. Investigation of this requirement determined that germination was triggered by exposure to red light of wavelength 660 nm, but inhibited by exposure to far red light of wavelength 730 nm. It is known that the red light is absorbed by the pigment **phytochrome** which then becomes converted to its physiologically active form.
- Chlorophyll formation does not take place in the absence of light. Seedlings deprived of light grow tall and straggly, with thin stems and undeveloped, yellow leaves. Such seedlings are described as **etiolated**.
- Light is essential for the process of photosynthesis. The green pigment, chlorophyll, present in the chloroplasts, absorbs red and blue wavelengths of the visible spectrum in the light-dependent stage, resulting in the production of ATP and reduced NADP for the light-independent stage.
- The duration and intensity of the light determines the rate at which photosynthesis occurs.
- Day length affects the initiation of flowering in some crop plants.

In order to maximise productivity, the crop needs to make best use of the available light during its growth. In many crops, it has been shown that there is a linear relationship between the growth rate and the amount of light intercepted. The more light absorbed, the higher the growth rate and the greater the dry matter produced. On a bright day, when the light intensity during the middle of the day is high, the leaves at the top of the crop receive more light than they can use, as light saturation is reached at relatively low light intensities. However, further down, the light intensity decreases. The leaves do not usually overlap completely, so receive some light through gaps in the canopy.

Availability of mineral ions

Mineral ions are taken up from the soil water by the roots. Most of the absorption occurs in the root hair region, where the thin cellulose cell walls of the root hair cells are freely permeable to the ions and where there is a large surface area over which uptake can occur. Uptake through the cell surface membrane into the cytoplasm and vacuole is affected by both the concentration gradient and the electrochemical gradient. Where the internal concentration of an ion is less than its concentration in the soil water, it diffuses into the root hair cell along the electrochemical gradient. In many cases, the uptake is a process of accumulation against the concentration gradient. Ions move from a lower concentration in the soil water to a higher concentration in the root cells, an active process requiring energy from respiration. There is evidence that ion uptake by the roots is selective. The relative concentrations of different ions inside the cells differs from their relative concentrations in the soil water. Cells take up those ions they require in preference to others.

Water

Water is taken up by the roots from the soil. Most of this uptake occurs in the root hair region, just behind the root tip, where the surface area available is increased by the presence of the root hairs. The water then passes across the cortex of the root, mostly via the **apoplast** route through the cell walls, into the xylem from where it is transported to the leaves.

Water is essential to plant growth. It is needed:
• to maintain the turgidity of the cells
• for the transport of soluble materials, such as ions and the products of photosynthesis, around the plant
• as the medium in which all metabolic reactions take place
• as a raw material in the synthesis of organic compounds, providing a source of electrons and hydrogen ions for the light-dependent stage of photosynthesis.

Water moves by diffusion from areas of high water potential (the soil) to areas of lower water potential (the leaves). Of the water which is taken up from the soil, only about 5 per cent is used by the plant, the rest being lost as water vapour from the leaves in **transpiration**. The water potential of the atmosphere is much lower than the water potential in the plant, so there is a tendency for water to be drawn up the plant to replace that which is continually being lost from the leaves.

Water uptake is high when the stomata are fully open in conditions of high light intensity during the middle of the day. It increases still further if the air temperature is high and if there are any air movements, as these factors increase the rate of transpiration. Uptake is reduced by low availability of water in the soil and by those factors, such as low temperature, high humidity and low light levels, which reduce the rate of transpiration. If water loss through transpiration exceeds water uptake by the roots, then **wilting** occurs. Often this situation is temporary and the plant recovers quickly when water is again available, but if the water shortage is prolonged it results in the death of the plant. Wilting causes the stomata in the leaves to close. In addition to reducing the loss of water vapour, wilting causes a reduction in the uptake of carbon dioxide for photosynthesis, thus reducing the growth of the plant.

The quantity of water required by a crop varies with the species and with the time of year or season in which it is grown. Water availability in different parts of the world determines which crops grow satisfactorily and a number of different agroclimatic zones have been defined, based on temperature and rainfall. In order to grow crops without irrigation, at least 50 cm of rain per year is needed. In many arid regions, irrigation schemes have enabled the utilisation of otherwise useless land for crop production. A good knowledge of the climate and of the requirements of particular crops at different stages of their growth have enabled farmers to irrigate at appropriate times, thus saving water and cutting down on expense. Even in temperate regions, where rainfall exceeds 50 cm per year, irrigation at certain periods of crop development can be beneficial. All crops require water for germination: some crops such as tomatoes and cucumbers require more water at the fruit development stage, whereas leafy crops benefit from watering during vegetative growth.

Maximising productivity

One of the ways in which crop producers can maximise productivity is to ensure a plentiful supply of mineral ions. In order to maintain fertility, the crop producer needs to apply fertiliser to the land. Different crops have different requirements for mineral ions, but all require adequate amounts of nitrogen, phosphorus and potassium if high yields are to be maintained. The development of the Haber process in 1908 for converting nitrogen gas into ammonia resulted in the production of large quantities of nitrogen-containing fertilisers. This in turn has contributed to the widespread use of inorganic fertilisers in preference to organic ones such as farmyard manure. Inorganic fertilisers are easy to apply in the required amounts and they are soluble so the ions are readily available to the plant roots, but over-application can lead to environmental problems such as eutrophication.

With reference to the nitrogen cycle how are nitrate ions made available to plants?

Nitrate

Nitrates are essential for the synthesis of amino acids and hence proteins, nucleic acids, pigment molecules and coenzymes. A nitrate deficiency shows first in the older leaves of the plant: they become yellowish-brown and fall off. The plants may eventually become spindly, pale coloured and have a low leaf to stem ratio. The deficiency can be remedied in young plants by supplying nitrate-containing fertiliser. Care should be taken to avoid over-application as this could result in weak growth and a decrease in carbohydrate production.

Nitrogen-containing fertilisers may supply nitrogen as nitrate ions (e.g. sodium nitrate) or as ammonium ions (e.g. ammonium phosphate, anhydrous ammonia or aqua ammonia). The nitrate ions can be readily absorbed by the plants. Some plants can absorb ammonium ions, but these ions are oxidised to nitrate ions by bacteria in the soil. The nitrogen-containing compounds in organic fertilisers such as farmyard manure and compost are insoluble and the material has to be broken down by microorganisms before there is any benefit to the crop. Both nitrate and ammonium ions stay in solution in the soil water and are not held around the soil particles by adsorption, so they are easily leached from the surface layers of the soil and can drain off agricultural land

PLANT GROWTH FOR CROP PRODUCTION

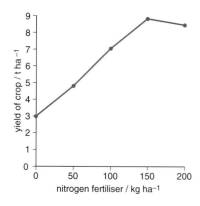

Figure 1.7 Effect of increasing nitrogen fertiliser on crop yield

Using the information given in Figure 1.7, calculate the percentage increase in yield given by the different levels of nitrogen fertiliser. What advice would you give to the crop grower about the level of nitrogen fertiliser to use?

into lakes and reservoirs. Particular concern has been shown about the consequences of this run-off, both in terms of eutrophication and in the levels of nitrates in drinking water.

Phosphate

Phosphates are associated with the transfer and storage of energy and are therefore vital for metabolic processes in plants. A supply of phosphate ions is essential in the early stages of growth and again during seed formation. A deficiency results in a reduced growth rate and poor root development. If the deficiency is severe, the plant becomes stunted. Phosphate deficiency is difficult to detect in its early stages because the leaves look healthy, but by the time growth appears stunted, it is often too late to remedy it. If seed formation is affected, any seeds produced are likely to have reduced vigour, thus carrying over the condition to subsequent crops.

The concentration of phosphate ions in the soil water is usually low because they readily form complexes with iron and aluminium ions, which are then adsorbed on to the surface of the clay particles. The ions are then `fixed' and unavailable to the plant roots. The presence of high levels of calcium ions in chalky or limed soils also decreases the availability of phosphate. Phosphate fertilisers usually consist of superphosphates or ammonium phosphate, both of which are soluble in water. Ammonium phosphate also supplies nitrogen in the form of ammonium ions and is a component of blended or compound fertilisers.

Potassium

Potassium ions are essential components of cell sap and are involved in the control of the water content of cells and the movement of water through the tissues. Potassium ions are also involved in the turgor changes in guard cells which bring about the opening and closing of the stomata. In addition, potassium is associated with the formation and translocation of carbohydrates and the activation of some enzymes. Plants vary in their requirements for potassium, but it is always needed in high concentration. A deficiency of potassium results in the margins of the leaves turning a bronze colour and beginning to die.

Fruit and leafy crops remove large quantities of potassium ions from soils, so in order to maintain yields potassium sulphate or potassium chloride is applied as a fertiliser. Potassium chloride is cheaper, but horticulturists prefer potassium sulphate as it is considered to give a better quality crop.

Different crops require different amounts of mineral nutrients, so the crop producer should be able to tailor the amount to the particular crop being grown. NPK fertilisers are commonly used and can have different ratios of constituents. A fertiliser described as 10:10:10 will have 10 per cent nitrogen, 10 per cent phosphate (in the form of P_2O_5 – phosphorus pentoxide) and 10 per cent potassium (in the form of K_2O – potash).

It is important to ensure that the fertiliser is available to the crop when the uptake of mineral nutrients by the plants is at its greatest, so that maximum benefit may be derived. Experiments and trials into the uptake of nitrogen by crops have resulted in recommendations to farmers as to the correct times at which to apply the fertilisers and in the correct quantities. This enables more effective utilisation of the nitrogen and less wastage. One of the most important factors is to guard against loss of fertiliser due to leaching, so careful consideration is given in the UK to applications made during the winter and the spring. When winter wheat is sown in the UK, no nitrogen fertiliser is applied until the following spring, but spring wheat is sown and fertiliser is applied straight away. If fertilisers are applied too early in the spring, there could be losses due to leaching, so applications to most cereal crops are made in late April or early May, when rapid growth is occurring.

Used wisely, inorganic fertilisers can help to maintain soil fertility and increase crop yields, but they are expensive as they have to be processed. The alternative strategy is to make use of organic material such as **farmyard manure**, **compost**, **animal slurries** or **sewage sludge**. In each of these alternatives, the essential mineral ions are not instantly available and there is always a high carbon to nitrogen ratio. Decomposition of the organic material by microorganisms in the soil has to take place. During this process, the carbon-containing compounds are used to provide energy for the microorganisms and carbon dioxide is released. Ammonium ions and other nitrogen-containing compounds are taken up rapidly as the populations of microorganisms increase. When the microorganisms die, the mineral ions are released into the soil and become available for uptake by the roots of the crop. The process of decomposition and release of mineral ions is slow, so it is much easier for crop producers to use inorganic fertilisers. The use of organic fertilisers does have advantages in that the material helps to maintain a good soil structure, improving aeration and drainage. Table 1.1 summarises the different types of organic material which may be used, together with some advantages and disadvantages of their use.

Table 1.1 *Organic fertilisers, their composition and use*

Type of material	Composition	Use
farmyard manure	urine and faeces from farm animals mixed with straw	spread on land, ploughed in; improves aeration and drainage; should be well-rotted and contain plenty of straw; difficult to handle; smelly
animal slurry	semi-liquid forms of faeces and urine from farm animals	similar to farmyard manure but more liquid; can be sprayed on to grassland or on to soil before planting; easier to handle but smelly
sewage sludge	by-product from sewage treatment works	cheap, plentiful, but may contain heavy metal ions; odourless product available from Wessex Water
compost	vegetable and animal wastes from household and garden	used by gardeners; best type is aerobic compost; blackish-brown in colour and crumbly; spent mushroom compost often used

Green manuring is a term which is used to describe a number of different practices including:
- the ploughing in of crop residues from vegetable crops, for example pea haulms, sugar beet leaves

- the ploughing in of a quick-growing crop such as mustard or rape
- leaving crop remains on the surface of the soil, with the subsequent crop planted through it by direct drilling.

These practices add organic material to the soil; decomposition takes place releasing nutrients slowly. Clover or field beans are often used as green manures as they possess nodules containing nitrogen-fixing bacteria. The use of quick-growing crops or the leaving of crop residues in the ground provides cover, preventing soil erosion and helping to control weeds. In addition, the plant cover reduces losses of mineral ions due to leaching.

Control of the environment

The growth of crops in glasshouses enables the producer to manipulate the environment to give optimum conditions for photosynthesis. The practice of using glasshouses, or greenhouses as they were formerly termed, has been in use since the 17th century, but it is only relatively recently that the control of the conditions has been so precise. It is now possible to use computers to monitor and adjust such conditions as light intensity, temperature, pH, mineral ion concentration and carbon dioxide concentration, maximising crop production and reducing labour costs. Such systems are expensive to install and to operate, but the increased yields and profits for the producer make it economically viable.

In many glasshouses, natural light can be supplemented by artificial lighting. The quantity of natural light trapped can be enhanced by carefully selecting the aspect, or siting, of the glasshouse and by using an asymmetrical roof shape. The most widely used form of artificial lighting is high-pressure mercury vapour lamps, which supply light of the correct wavelengths for photosynthesis. Domestic light bulbs are inadequate and sodium vapour lamps, although more efficient, do not give the correct wavelengths. As any supplementary lighting is expensive, it is usually only used when plants are small and they can be grown close together on benches in the glasshouse.

In order to maintain temperatures above the minimum for a particular crop, some form of heating system is used. The system is thermostatically controlled and the heaters may provide hot air from oil, gas or electricity, involve boilers supplying hot water to a series of radiators, or warm the soil either by hot water pipes or electric elements. When the temperature becomes high, it may be necessary to use some form of ventilation combined with shading screens during the summer when high light intensity can also increase temperatures. Ventilation systems are installed in commercial glasshouses. These consist of vents in the roof, which can be opened allowing the hot air to escape, while vents near the ground permit the entry of cooler air. One of the consequences of using these ventilation systems is that they can alter the humidity of the atmosphere inside the glasshouse, which, in turn, affects the transpiration rate of the crop. Increase in temperature and increase in air movement will increase the rate of transpiration, so it is desirable to use some means of increasing the humidity, such as misting, while the vents are open.

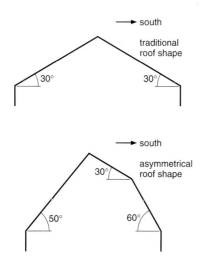

Figure 1.8 Aspect and roof shape in glasshouses

As mentioned earlier, the carbon dioxide concentration in the atmosphere is around 0.03 per cent and it has been shown that increasing this concentration can increase the rate of photosynthesis. In glasshouse cultivation, the carbon dioxide could be used up quite rapidly if the temperature and the light intensity are high. Ventilation enables supplies of carbon dioxide to be replenished from the atmosphere, but concentrations can be increased by using paraffin burners, dry ice or gas from a cylinder. If higher concentrations of carbon dioxide are required, they are used during the day, when it is light and it is necessary to keep the vents closed to prevent the gas escaping into the atmosphere. The conditions needed and the expense involved make such enrichment of the carbon dioxide concentration of limited value to most crops, but it has been shown to be economical for tomatoes and cucumbers, where concentrations of 0.06 per cent are commonly used.

The control of water and mineral ions can be achieved by growing glasshouse crops using **hydroponics**, or the **nutrient film technique (NFT)**. This technique was developed in the 1970s at the Glasshouse Crops Research Institute in the UK and avoids the use of soil, thus eliminating problems such as poor soil structure, variable mineral ion content, pests and soil-borne diseases. The plants are grown in troughs, often with their roots embedded in rock wool, or some other inert material. They are supplied with a nutrient solution, containing the correct balance of mineral ions, which is pumped into the troughs and allowed to flow around the roots before being collected in a tank and re-circulated. As the plants grow, they are supported if necessary by strings or wires suspended from the framework of the glasshouse. The composition of the nutrient solution can be monitored continually and the mineral ion concentration, pH, oxygen content and temperature adjusted as required. This system is widely used in growing tomatoes and cucumbers.

Figure 1.9 Hydroponic system for the cultivation of tomatoes.

The ability to control the duration of the day length, or photoperiod, in a glasshouse has affected the commercial production of flowers. With the use of artificial lighting during the winter and shading during the summer, it is possible to manipulate the amount of light received by the plants and so prevent or initiate the flowering process as required. Chrysanthemums are short-day plants, which are induced to flower when the day length is less than 14 hours, so in temperate regions they normally flower during the autumn and the winter. In order to maintain a supply of flowers for sale throughout the year, the plants are kept in the vegetative state until flowers are required. This is achieved by lengthening the photoperiod, mimicking long days, or giving a short period of illumination during the middle of the night. Eight or ten weeks before flowering is required, the plants are transferred to conditions of the correct day length, thus initiating the production of flower buds. A more detailed explanation of the photoperiodic effect and the initiation of flowering is given in Chapter 3 of *Systems and their Maintenance*.

Interaction between crop plants and other organisms

Interactions between crop plants and other organisms range from competition with other plants for resources, such as nutrients and light, to infections and infestations caused by pathogenic microorganisms and insect pests. Competition may be **intra-specific**, that is between the crop plants themselves, or **inter-specific** between the crop plants and weed species.

Spacing and its effect on crop yield

The density at which a crop is sown, or planted, has an important effect on the yield. The crop producer needs to ensure that best use is made of the available land in order to achieve a high yield. If the crop is too densely planted, there will be intra-specific competition for the available light and nutrients, which may result in small seeds or fruit. If the crop cover is too sparse, there will be increased inter-specific competition from weed species. A sparse crop can also lead to excessive branching or tillering, where the seeds or fruit produced may be small. In addition, due to the later time of flowering of the tillers, ripening of the crop may occur over a longer period of time, resulting in unripe seed being harvested with the ripe crop. The aim is to grow the crop at such a density that the maximum yield is obtained for the least number of seeds or plants.

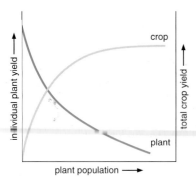

Figure 2.1 Plant and crop yield in response to increasing plant density

If a crop plant is grown on its own, with no competition and unlimited resources, its yield gives an indication of the maximum yield possible per plant for that crop. As the density of the crop is increased, competition between individual plants increases. The yield per individual plant decreases but the crop yield will increase. Increase in density makes more efficient use of the available resources, particularly the interception of light. However, as the plant density increases further, the increase in yield is not as great and eventually a plateau is reached.

At higher plant densities, there is greater vegetative growth and less reproductive and storage tissue is produced. The nature of the crop, whether it is harvested for the roots or for the fruits, has to be taken into consideration when determining the density at which the crop is sown.

In order to achieve a suitable density of plants, it is necessary to sow an adequate number of seeds. This **seed-rate**, expressed as the mass of seeds needed in kg per hectare, varies according to the crop and how well the plants become established subsequent to germination (the **field establishment**). This, in turn, depends on the weather and soil conditions at planting. For example, winter cereals sown early in the autumn on good seedbeds have a higher percentage field establishment than those sown later in cold, wet soil. In order to calculate an appropriate seed-rate for a winter cereal crop, the producer needs to know:

- the **thousand grain weight (TGW)** for the cereal, which is the mass of 1000 grains in g and is a measure of the size of the grain
- the required density of the established plants per m², known as the **target population**
- the expected percentage field establishment.

The seed-rate, in kg per hectare, can then be calculated using the following equation:

$$\text{seed-rate} = \frac{\text{target population} \times \text{TGW}}{\text{expected field establishment}}$$

For example, if:

target population	= 250 plants m^{-2}
TGW	= 48 g
expected percentage field establishment	= 75%

then

seed-rate	= 250 × 48 ÷ 75
	= 160 kg ha^{-1}

The yield of grain from a cereal crop depends on the number of ears per plant, the number of grains in each ear and the mass of the individual grains. The number of ears produced by each plant depends on the number of side shoots, or tillers, that arise from axillary buds on the stem. Each of these tillers can produce an ear, but cereal crops differ in their capacity to produce tillers. Winter wheat and winter barley can produce tillers in large numbers, especially if the crop is planted early in the autumn on a good seed bed. In practice, most tillers die back, leaving 1.5 to 2 tillers per plant for winter wheat and 2 tillers per plant for winter barley. Under these circumstances, less seed may be required to produce a high yield. Spring-sown cereals produce fewer tillers, so a higher seed-rate may be necessary to achieve a similar yield.

When planting arable crops, some consideration should be given to the spacing of the seeds and the width of the rows. Most crops are sown mechanically and the drills used can be set accurately to deliver the seed at the most suitable spacing. As mentioned before, the producer needs to achieve a plant density that will give a good yield, and the spacing of the seeds and rows should permit cultivation techniques, such as spraying, weeding and harvesting, without causing damage to the crop. The row width for most root crops and maize is determined by the harvesting machinery. For a root crop such as sugar beet, where the seeds supplied are between 3.40 and 4.75 mm in diameter, it is recommended that the grower should aim for a plant density of 75 000 plants per hectare. The inter-seed spacing is normally 175 mm and the rows should not be more than 500 mm apart.

Intercropping, sometimes referred to as **mixed cropping**, involves the growing of two or more crops together in the same area at the same time. There are several different ways in which this type of cropping is managed. These include:

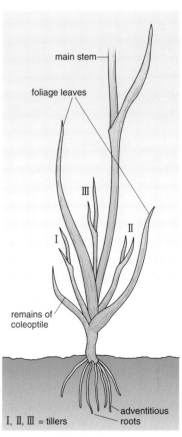

Figure 2.2 Tiller development

- planting a slow-growing crop, such as parsnips, with a faster-growing one, such as radishes or lettuces: the radishes or lettuces will be ready for harvesting before the parsnips need the space
- the cultivation of low-growing crops beneath taller ones: lettuces may be grown beneath sweetcorn or winter brassicas (sometimes known as **undercropping**)
- the use of the space between the rows of a maturing crop as a seedbed for the production of young plants.

Intercropping has been more common in the tropics and Asia than it has been in Britain, but it is becoming increasingly common in gardens and smallholdings, where space is limited. It has a number of advantages to the grower, such as:

- making more efficient use of the available land
- making more efficient use of resources such as light by having a denser plant cover
- more mature, or taller, plants providing protection for smaller, developing ones
- reducing the competition from weeds
- reducing the possibility of soil erosion as the land is covered with plants for longer
- allowing greater profits for the grower, who benefits from the income from two crops instead of one.

In addition, if a leguminous crop is included, there will be benefits from nitrogen fixation.

The main disadvantage of such systems is that the crops may mature at different times, creating problems with harvesting. It is usually not possible to harvest the crops mechanically, so collecting each crop could be more labour-intensive.

Figure 2.3 Intercropping

Competition from weeds

A **weed** can be defined as a plant that grows in a situation where it is not wanted. This definition has been modified to include plants which are not deliberately cultivated by humans but which are able to grow in situations disturbed by them. A field cultivated to provide ideal conditions for the growth of an arable crop also provides ideal conditions for the germination and growth of the seeds of weed species. These seeds may be wind-dispersed, or they may come from the machinery used to prepare the ground, from farmyard manure, from hay put out for livestock or from passing animals and birds. Sometimes, a crop may be contaminated by the remains of a previous crop. These contaminants are known as **volunteers** and may be derived from seeds or organs of vegetative reproduction which have been left in the soil.

The control of weeds is of great importance to crop producers because weeds compete with the crop for the available resources, such as light, water and mineral nutrients, causing a potential reduction in the yield of the crop. Most of the competition between the crop and the weeds occurs in the early stages of growth, but weeds can affect crop production and farming practices in other ways. They may, for example:

- produce chemical compounds from their roots or leaves which inhibit the germination of the seeds of other plants (**allelopathy**)
- contaminate seed crops with their seeds; a particular problem is contamination of the pea crop, used for canning and freezing, with the fruits of black nightshade (*Solanum nigrum*)
- be hosts for the diseases and pests of the crop plants, e.g. fat hen (*Chenopodium album*) and some leguminous weeds can be hosts for the black bean aphid (*Aphis fabae*) a pest of field beans; wild members of the Cruciferae can host *Plasmodiophora brassicae*, the organism which causes club root in cabbages
- be poisonous to grazing animals and contaminate hay and silage
- have spines or thorns which could harm grazing animals
- be unpalatable to animals, nutritionally poor or taint animal products, e.g. wild onion imparts an unpleasant flavour to milk and meat
- interfere with farm machinery especially during harvesting and later cultivations, e.g. knotgrass (*Polygonum aviculare*)
- block drainage channels and irrigation ditches, e.g. water hyacinth (*Eichornia crassipes*)
- become established in places, such as road verges and railway embankments, where they provide a reservoir of seeds which can colonise agricultural land when the opportunity arises.

Before considering the characteristics which enable weeds to be so successful in competing with crop plants, it is worthwhile noting that they do have some beneficial effects. Because they are able to colonise bare soil rapidly, they provide a valuable cover of vegetation which can reduce soil erosion and prevent the leaching of valuable mineral nutrients. In addition, the growth of weeds in hedgerows and at the edges of fields provides habitats for wildlife, especially for insects which may be crop pollinators. Some weed species, closely related to crop plants, could possibly provide a source of genetic material for the breeding of new varieties of the crop species.

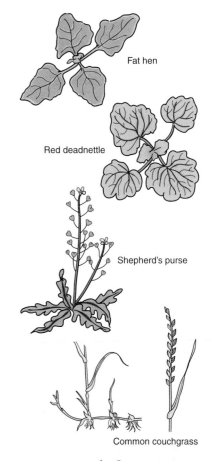

Figure 2.4 Some common weeds

Characteristics of weeds

Many weed species are ephemerals, having very short life cycles. They can germinate rapidly, their seedlings grow quickly and flower production, followed by seed production, occurs after a relatively short period of vegetative growth. Thus many of these plants can complete more than one life cycle in a growing season. When conditions are favourable for growth, there is a high output of seeds, but even when conditions are poor there is some seed production. In some species, seed production can continue over a long period of time and seeds may remain viable in the soil for long periods. Other species are **annuals**. They may be winter annuals, with seeds germinating in the autumn. The seedlings which develop may grow rapidly, flower and produce seed before the winter, or they may remain in a vegetative state during the winter and flower the following spring. The seeds of the summer annuals germinate in the spring, produce flowers and seeds during the summer before dying. **Perennial** weeds produce seeds and also have organs of perennation, such as underground stems (rhizomes) in couch grass (*Agropyron repens*) and tap roots in dandelions (*Taraxacum officinale*). These perennating organs have food reserves and are very difficult to eradicate from cultivated land, but they enable successful competition with other plants.

For all weed species, the production of large quantities of seed is of paramount importance for survival. For example, groundsel (*Senecio vulgaris*) can produce between 1000 and 2000 fruits per plant, shepherd's purse (*Capsella bursa-pastoris*) produces up to 4000 seeds per plant and the common poppy (*Papaver rhoeas*), with its large round seedcases, about 19 500 per plant. Most seeds are small and light, often with devices aiding dispersal. All these examples are annual or ephemeral weeds, but perennial weed species are just as prolific, despite having other means of spreading. Many weed species are self-compatible, eliminating the need for cross pollination.

Variable seed dormancy is a characteristic which has contributed to the success of weed species. Seeds show three types of dormancy:
- **innate**, which is genetically determined: seeds will not germinate for a period after they are shed from the plant
- **induced**, which is caused by the presence of a specific condition, such as high concentrations of carbon dioxide: buried seeds do not germinate, but when brought to the surface by ploughing, the condition is removed and germination can occur
- **enforced**, which occurs when environmental conditions are unsuitable: some seeds require light and do not germinate if they are buried.

A large number of weed seeds show innate dormancy, but some show the other types and germinate readily when brought to the surface by cultivation techniques. In contrast with crop seeds, weed seeds can survive for long periods of time if buried in the soil and a high percentage of these seeds are capable of germination. It has been demonstrated that most cultivated arable land contains large numbers of weed seeds, forming a seed bank ready to take advantage of favourable conditions as they occur.

Patterns of germination

Some weeds show rapid and **synchronous** germination, similar to crop plants, while others show germination over a longer period and there are some which show **intermittent** germination. All these strategies have their advantages and there can be variation in the conditions needed within the same genus. For example, non-weed species of the genus *Chenopodium* have a much narrower range of conditions for successful germination than the weed species fat hen. In general, weed species are more tolerant of short-term variations in the physical environment, such as extremes of temperature and high evaporation rates, than are their non-weed relatives. As mentioned above, seed dormancy plays an important role and is a survival mechanism. As the duration of dormancy can be many years, there is always the potential for germination of some weed seeds to occur as soon as the soil is disturbed.

Many common weed species can germinate at any time of year, whereas in others germination is restricted to certain seasons. Table 2.1 shows the germination periods of some common annual weeds and it is clear that there is no season of the year when some weed germination does not occur.

Table 2.1 *Germination periods of some common annual weeds*

Species	J	F	M	A	M	J	J	A	S	O	N	D
Avena fatua (Wild oat)		●	●	●	●	●		●				
Capsella bursa-pastoris (Shepherd's purse)	●	●	●	●	●	●	●	●	●	●	●	●
Polygonum aviculare (Knot grass)		●	●	●	●	●						
Polygonum convolvulus (Black bindweed)				●	●	●						
Stellaria media (Common chickweed)	●	●	●	●	●	●	●	●	●	●	●	●
Galium aparine (Cleavers)	●	●	●							●	●	●
Plantago major (Greater plantain)					●	●	●	●				
Solanum nigrum (Black nightshade)					●	●	●	●				
Senecio vulgaris (Groundsel)	●	●	●	●	●	●	●	●	●	●	●	●
Papaver sp. (Common poppy)		●	●	●	●	●	●	●	●	●	●	

When competing with crop plants, weed seeds that are able to germinate and start their growth before the crop seedlings become established have an advantage. It is likely that the weed will overgrow the crop, which will be deprived of light, water and nutrients and there is little doubt that the crop yield suffers if there is a heavy infestation of weeds.

In the early stages of growth, the crop could be deprived of water as the weed cover may prevent rain from penetrating through into the soil. Development of root systems is important in plants and weed species have been shown to develop just as extensive a root system as the crop plants. Some weed species are known to be particularly efficient at taking up specific mineral nutrients. For example, fat hen (*Chenopodium album*) takes up potassium ions readily and experiments have shown that when couch grass (*Agropyron repens*) is a weed amongst maize, it absorbs most of its mineral requirements early in its growth so that the maize, which starts its major period of growth later, suffers from reduced nutrient availability. Black grass (*Alopecurus mysuroides*) competes with cereal crops for nitrogen early in their development.

Experiment A

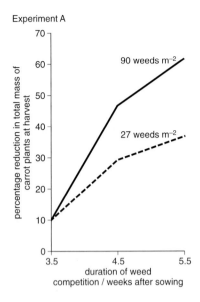

Experiment B

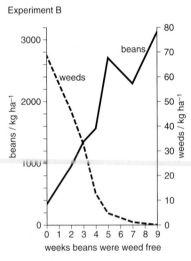

Figure 2.5 Results of experiments on effects of competition from weeds on crop yield. Experiment A results show that the density of weeds in a carrot crop will affect the yield at harvest. Experiment B results show that as the length of time the beans were free from weeds increased, so did the yield

The greatest effect of competition from weeds occurs during the early stages of crop growth, the critical period being the first 4 to 6 weeks, so it is important to keep the crop weed-free for this period so that the crop plants can become established.

Control of weeds

For centuries, the removal of weeds from crops was done by hand or by ploughing, but in the 1940s, the first chemical herbicides were developed. At present, **herbicides** are classified in a number of ways:

- **contact** – acting only on the regions of the plant with which they come into contact
- **translocated** – taken up and moved around the plant so that distant sites, such as the roots, may be affected
- **selective** – growth regulators, which are absorbed by the leaves and translocated around the plant, killing some groups of plants, but leaving others unaffected
- **non-selective** – applied to kill all vegetation before planting a crop
- **residual** – usually soil-acting; applied directly to the soil, where they remain in the top 2 cm, killing off germinating seeds, but not affecting deeper-rooted species.

The choice of herbicide and the timing of its application depend on the nature of the crop, the method of cultivation and the weed species. **Pre-sowing** treatment involves application of the herbicide before the seed is sown in order to kill off the weed species. A contact herbicide, such as paraquat or diquat, could be used to kill off the aerial parts of the weeds, prior to sowing the crop. This method is often used where a crop is to be sown without ploughing the land beforehand, a process known as direct-drilling. Such herbicides act quickly, inhibiting photosynthesis in the weeds by destroying the thylakoid membranes of the chloroplasts. These herbicides break down rapidly into harmless products, leaving the soil uncontaminated. A soil-acting herbicide, such as Linuron, which also inhibits photosynthesis, could be used at this stage. This herbicide binds to the soil particles and kills the weed seedlings as they emerge. **Pre-emergent** herbicides are applied after the seed is sown but before the crop emerges.

Post-emergent herbicides need to be selective in their action as they are applied to both the crop and the weed. For some time, herbicides such as 2,4-D (2,4-dichlorophenoxyacetic acid), fosamine, dicamba and picloram have been used to kill off broad-leaved (dicotyledonous) weeds, leaving the narrow-leaved (monocotyledonous) plants unaffected. These herbicides are particularly useful in the weed control of cereal crops. 2,4-D causes abnormal growth: cell division and elongation in roots is inhibited and metabolic processes, such as photosynthesis and respiration, are disrupted. Nowadays, there is a very wide range of selective herbicides available so that it is possible to control monocotyledonous weeds in a cereal crop (wild oat, *Avena* spp., can be eradicated from wheat), monocotyledonous weeds in a broad-leaved crop (wild oat in sugar beet) and broad-leaved weeds in a broad-leaved crop (fat hen in kale and rape). It should be noted that some selective herbicides may have an effect on the crop if enough chemical is retained in the plant or the crop is under stress.

Glyphosate is a broad spectrum herbicide, with a structure similar to the amino acid glycine, and has been found to be toxic to many types of plants. It is readily translocated around the plant and is broken down to harmless products by microorganisms in the soil. It has been shown to have a low toxicity to mammals.

It is possible to control weeds without using herbicides, but the procedures involved are more time-consuming and not always as effective. **Cultivation**, or **secondary tillage**, involves hoeing, discing and harrowing, usually with a tractor. These methods will kill off weed species that are growing and are particularly effective against perennial weeds. Ploughing is the most effective cultivation method as it buries the weed seeds. Inter-row cultivators are used for crops, such as maize, sugar beet and potatoes, which are grown in wide rows. The use of flame-weeders, where weeds are exposed to temperatures in excess of 100 °C, is expensive as the apparatus uses propane fuel. The high temperature dehydrates the tissues of the weeds and coagulates the proteins. This method is not economical to use on a large scale, but is cheaper than hand-weeding. Good farming practice, such as only sowing clean, certified seed, burning straw infected with weeds and intercropping can all contribute to fewer weeds in the crop. **Crop rotation** can prevent the build up of certain weed species and if cereal crops are alternated with broad-leaved crops, then this allows the use of different herbicides. It is also good practice to remove weeds from the edges of fields, either from the hedgerows or from the headlands surrounding the crop, so that the seeds do not spread on to the cultivated area.

Band spraying is a method of weed control, which combines the use of herbicides with cultivation. It is used on crops such as sugar beet, which are grown in rows. The crop is sprayed with herbicide along the rows and hoed between the rows. The application of the herbicide is made using a shield to retain the spray, so that only the rows receive the herbicide. This method of weed control reduces the use of chemicals by up to two thirds.

Where crop production is on a smaller scale, in gardens and smallholdings, weeds are often removed manually. In the production of cereals for seed, weeds such as wild oats (*Avena fatua*) are still pulled manually. The method is referred to as **hand rogueing** and may involve the use of a glyphosate-impregnated glove (chemical rogueing). More use is made of the practice of intercropping and **mulching**, involving the use of peat, bark chippings, straw, manure or black plastic sheeting on the soil surface around the crop plants. Both these methods have other advantages, especially in reducing the amount of water evaporated from the soil surface, but the use of peat as a mulch is being discouraged due to the rapid loss of unique peatland habitats.

The preparation of a 'false' seedbed is sometimes used in weed control. The land is prepared as if for sowing, but then left for a period, during which the

Figure 2.6 (top): modern weed control – band spraying crops with herbicide; (bottom) the old method – manual hoeing

Figure 2.7 Black plastic sheeting can be used in weed control. The sheeting is cut so that crop plants protrude through, but any weeds between them are deprived of light and die.

weed seeds germinate. The land is then harrowed mechanically or sprayed to kill off the weed seedlings and the true seedbed prepared.

Biological control of weeds involves the introduction of organisms which feed on the weed species but not on the crop. This type of control has been effective in Australia, where the insect *Cactoblastis* was introduced so that its larvae could feed on the prickly pear.

Fungal disease

Fungal diseases result in loss of yield, depriving the crop plants of nutrients and often causing breakdown of tissues resulting in the death of the plants. Such diseases account for losses of up to 15 per cent of crops world wide. Fungal pathogens of crops can spread very rapidly, so the best methods of control involve the prevention of infection by protecting the crop and eliminating the sources of infection.

Fungal diseases in crops may be air-, soil- or seed-borne and many may show two or more phases of attack. Most fungal plant pathogens rapidly produce masses of asexual, air-borne spores, which can be dispersed over long distances in a short period of time, so air-borne infections quickly become established in a crop, attacking the leaves and stems. Nearly all the plants in a crop can be affected to the same extent at the same time. In order to control outbreaks of such diseases, it is necessary to know the life cycle of the pathogen and gain an understanding of the conditions in which it flourishes. Many air-borne pathogens thrive in warm, humid conditions and are often very specific to particular host crops. Soil-borne diseases, due to fungal pathogens present in the soil, may be difficult to diagnose as the main symptoms occur at the base of the stem and the roots. The affected plants may show stunted growth or wilting and their distribution in the crop may be patchy. Seed-borne diseases, due to infection by the pathogen of the reproductive structures of the plant, may affect young crop seedlings or may not show until the crop produces flowers.

The cycle of a fungal infection begins with the penetration of the host plant by the fungal pathogen. Once this penetration has been achieved, either through the epidermal tissues of the host or via the stomata on the aerial parts, the pathogen can grow and spread rapidly, absorbing water, mineral nutrients and the products of photosynthesis of the host. At this stage, the host plant may show signs, or symptoms, of infection, such as yellowing of the leaves or wilting. Asexual reproductive structures are usually produced by the pathogen soon after infection. Depending on their nature, these structures can cause the infection to spread further within the host plant or bring about infection of neighbouring plants. Some fungal pathogens are **obligate parasites**, only able to survive on living host plants, but others are **facultative parasites**, which infect a living host and then live as a saprobiont on its dead remains. Asexual reproduction usually results in the production of large numbers of spores, so that the spread of an infection is rapid. Sexual reproduction, when it occurs, is often a response to adverse conditions, but produces resistant spores, capable of germination when conditions are favourable.

Pythium debaryanum is a soil-borne fungal pathogen of a large number of crop plants and causes a condition known as 'damping-off' in young seedlings. The fungus belongs to the Zygomycota and has a mycelium lacking cross-walls (coenocytic). It is commonly present in soil where it feeds saprobiontically on plant remains. It can infect young seedlings, either from spores or mycelia in the soil, or from spores produced on neighbouring plants. Once penetration of the seedling has occurred, the fine hyphae branch and spread through the tissues absorbing nutrients and metabolites. The hyphae grow mostly between the cells of the host, but some penetrate the cells. In the early stages of the disease, the mycelium is restricted to the hypocotyl region of the stem, just above the soil level. The cells in this region of the host plant die, causing the stem to collapse. The hyphae may then grow into the vascular tissues, absorbing water, mineral ions and the products of photosynthesis.

Asexual reproductive structures are produced soon after infection and occur on aerial hyphae which grow out from the host. The tip of an aerial hypha swells up and develops into a zoosporangium, which is then separated by a septum, or cross-wall. If the atmosphere is dry, this sporangium, when it becomes detached, behaves like a spore and can be dispersed in air currents. When it germinates, a hypha is produced, which can infect a new plant, thus enabling reproduction to take place without the formation of zoospores when there is insufficient water for them to swim to new hosts. In damp conditions, the sporangium produces a vesicle into which the contents of the sporangium move. Here, zoospore formation takes place, each zoospore being kidney-shaped with two flagella attached laterally. When the vesicle breaks open, the zoospores are released and are able to swim away in the water films on the plants or around the soil particles. After a while, the zoospores stop swimming, become rounded in shape and develop walls. Under suitable conditions, these spores can germinate, producing fine hyphae, which are capable of penetrating a new host plant.

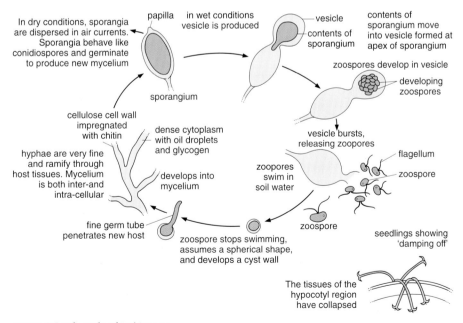

Figure 2.8 Life cycle of Pythium *sp.*

Sexual reproduction occurs within the tissues of the host plant, when the food material is nearly all used up. Thick-walled zygotes, called oospores, are produced, and these contain large numbers of oil globules, forming a food reserve. Oospores are only released on decay of the host plant tissue. They are resistant to desiccation and germination only takes place in the presence of moisture, so this stage in the life cycle enables the survival of adverse conditions.

Infections due to *Pythium* spp. can spread very rapidly, particularly in the warm, humid conditions found in glasshouses. Infections are most severe when moisture levels are high, as this favours growth of the fungus and the production of zoospores. As the disease can be soil-borne, it is important to sterilise the soil in the glasshouse, using heat or steam. Seeds can be treated with organomercury compounds before sowing, but care is needed as these compounds can be toxic to other organisms. Some research into the addition of microorganisms, antagonistic to *Pythium*, to the soil in glasshouses, has been carried out in an attempt to find a biological control for this pathogen, but this would not be practicable on a large scale. It has also been suggested that the addition of chitin to the soil will encourage the growth of a wide range of saprobiontic microorganisms, some of which may be antagonistic to the pathogen. However, the most widely-used methods of control involve the use of fungicides and the avoidance of over-crowding of the seedlings in the early stages of growth.

Control of fungal diseases

Fungal diseases in crops may be controlled by means of a combination of methods, depending on the nature of the pathogen and whether it is air-, soil- or seed-borne. Air-borne pathogens are usually controlled by planting resistant varieties of crop plants or by the use of chemicals, known as **fungicides**, which are toxic to the fungus but not to the crop plant. If resistant varieties of a crop do not exist, then producers are advised to grow several varieties of a crop, rather than planting the same variety year after year. Treatment of seeds with fungicide is not usually as effective in controlling air-borne infections as spraying the leaves and stems. It is advisable to use the fungicide at the first signs of disease, particularly if the environmental conditions are favourable for the reproduction and spread of the fungus. In many cases, crop producers spray susceptible crops if there is an outbreak of the disease in the neighbourhood. In the case of potato blight, caused by the fungus *Phytophthora infestans*, the conditions leading to its onset can be monitored. If the conditions are favourable, a blight warning is issued by ADAS (Agricultural Development and Advisory Service) and all potato crops will be sprayed. Once the first blight warning has been received, the crop needs to be sprayed regularly with fungicide.

Soil-borne diseases are extremely difficult to treat and to get rid of, but their effects on crop yield may be reduced by applying fertilisers, draining soil and the planting of resistant varieties. It is difficult to use chemicals: they are expensive and not always very effective. Where a particular soil-borne disease is persistent, it is usual to switch to a crop which is not affected.

Seed-borne diseases may be controlled by the use of healthy, certified seed or by the use of fungicidal seed dressings, such as benomyl, carboxin and thiram. Thiram is a **protectant** fungicide, which kills off fungal spores and mycelium on the outside of the plant. Such fungicides do not enter the tissues and are not translocated around the plant. They kill the fungus by disrupting enzyme systems and protein structures. Benomyl and carboxin are **systemic** fungicides, which are taken up into the plant and translocated to every part. When used as seed dressings, they are taken into the young seedlings soon after germination, giving protection to the whole plant as it grows and develops. Benomyl is known to interfere with microtubule formation in dividing cells, so disrupting the process of mitosis and inhibiting growth.

Both protectant and systemic fungicides may be used as foliar sprays on crops. The effectiveness of the protectant fungicides depends on the type of compound, its persistence and the timing of its application to the crop. As these fungicides remain on the outside of the plants, they may be washed off in the rain and re-application may be necessary to give protection against the disease. The spraying of large areas of crops with protectant fungicide is usually only undertaken following warnings of local incidence of a disease and careful consideration is given to the weather conditions. Systemic fungicides, applied as foliar treatments, can kill off fungal mycelium which may be present in the plants. Systemic fungicides give longer-lasting protection from disease than protectants.

Fungi can become resistant to a fungicide, though this is more common with systemic fungicides than protectants. Resistance develops most quickly with fungi that produce vast numbers of spores, with a greater capacity for variability. With this in mind, crop producers are recommended to use fungicides sparingly and to rotate their use, so that the same one is not used year after year. Mixtures of different fungicides have been tried and this approach does seem to avoid the establishment of resistant strains, as well as providing protection against a number of different fungal diseases.

Insect pests of crops

Insect pests cause considerable damage to crop plants both directly, by destroying plant tissue, and indirectly, either by reducing crop quality during storage or by the transmission of pathogens. The damage is caused by the feeding habits of both the larvae and the adults. The larvae of such groups as Diptera (flies) and Lepidoptera (butterflies and moths) and the adult Orthoptera, all have chewing mouthparts and cause damage to leaf tissue, thereby reducing photosynthesis. Insects with sucking or piercing mouthparts, such as Hemiptera (aphids), can cause a reduction in plant vigour by removing sap which contains the soluble products of photosynthesis. In addition, if a plant is infected with a virus, infected sap could be taken up by a feeding insect. The virus particles can multiply inside the aphid and then be transmitted to uninfected plants during subsequent feeding. Barley Yellow Dwarf Virus (BYDV) is spread in this way. Virus diseases, once contracted by a plant, cannot be controlled or cured, so the strategy is to prevent infection. In the case of BYDV, any growing plants that may carry the infection are removed and aphids are controlled in the autumn.

INTERACTION BETWEEN CROP PLANTS AND OTHER ORGANISMS

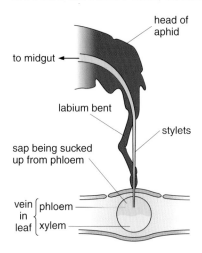

Figure 2.9 Aphid feeding on plant leaf

The **black bean aphid** (*Aphis fabae*) is a serious pest of both field beans and sugar beet. In field beans, dense colonies of aphids build up on stems and at growing points. The pods of the beans become distorted and sticky with honeydew. In crops of sugar beet, leaves are damaged due to direct feeding: the leaves turn brown and become distorted. In both crops, aphid infestations may result in the transmission of virus diseases. Colonies of aphids build up rapidly in hot, dry weather.

The life cycle of the black bean aphid shows incomplete metamorphosis, where eggs hatch into miniature adults, or nymphs, which undergo several moults, or ecdyses, before the adult emerges from the final moult. There is no larval stage and no pupa. Eggs of the black bean aphid overwinter on spindle trees. These eggs are black with thick outer cases and can withstand extremes of temperature. The eggs hatch out in March into wingless female nymphs, which feed on the young shoots and leaves of the spindle trees. At this stage, no males are produced and when the wingless females mature, they give rise to daughter nymphs by parthenogenesis, a process which involves no fertilisation. Some of these daughter nymphs develop into winged females, which migrate to food plants such as field beans. Although not strong fliers, these winged females can be carried over long distances in air currents. On reaching a food plant, the winged females give rise to wingless daughters by parthenogenesis. If the weather is warm, these daughters mature in about 10 days and begin to reproduce. Successive generations of winged and wingless females are produced throughout the summer, resulting in enormous numbers of aphids. The wingless forms remain and feed on the bean plants on which they are born, but the winged forms can migrate to new food plants. The aphids are preyed upon by ladybirds, lacewings and birds and are also killed off by cold weather, but infections can be severe. Towards the end of the summer, winged males and females are produced and both sexes migrate to the spindle trees. The winged females produce a generation of wingless daughters, which mature and then mate with the males, producing eggs.

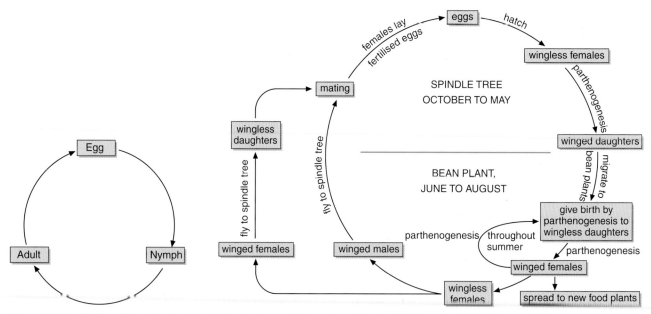

Figure 2.10 (a) Outline life cycle of an insect showing incomplete metamorphosis; (b) Life cycle of black bean aphid

INTERACTION BETWEEN CROP PLANTS AND OTHER ORGANISMS

Control of insect pests

Insect pests can be controlled in three ways:

- chemical control involving the use of insecticides
- biological control
- integrated pest management.

There are many types of insecticides available to the crop producer, but all of them involve some hazards and risks. Insecticides can either be absorbed directly by the insects, usually through their cuticle (**contact** insecticides) or they are taken up by plants so that their sap becomes toxic to the feeding insects (**systemic** insecticides). The organochlorines and pyrethroids are contact insecticides, whereas the organophosphates and carbamides are systemic. When considering the use of an insecticide, the crop producer needs to assess several factors, including:

- the nature of the insect pest
- toxicity to the insect pest
- toxicity to other organisms
- persistence in the environment.

The effects and uses of some of the major groups of compounds used as insecticides are summarised in Table 2.2.

Table 2.2 *Different types of insecticides*

Type of insecticide	Examples	Mode of action
organochlorines (chlorinated hydrocarbons)	lindane, metoxychlor (all approvals for DDT, aldrin and dieldrin were revoked in 1989)	used as powders, seed dressings, emulsions; absorbed directly by insects; act by interfering with nerve transmission; kill wide range of insects; persistent; accumulate in food chains
pyrethroids	rotenone, nicotine, extracts from flowers of *Pyrethrum*; synthetic: cypermethrin, deltamethrin	contact and stomach poisons, absorbed through insect cuticle or ingested; cause paralysis by interfering with nerve transmission; highly toxic to bees and fish; short-lived, breakdown rapid
organophosphates	malathion, parathion, diazinon, chlorpyrifos-methyl, chlorfenvinphos	more soluble in water than organochlorines; taken up by plants so systemic; effective against aphids; act by inhibiting cholinesterase at nerve synapses; more poisonous than organochlorines; poisonous to humans
carbamides	carbaryl, primicarb	similar to organophosphates, but more persistent; less persistent than organochlorines; carbaryl effective against caterpillars; primicarb used as a systemic to kill aphids

There are obvious problems with the use of insecticides to control aphids, as it is difficult to avoid killing off other useful insects such as their natural predators, the ladybirds and lacewings, or pollinating insects such as bees. Farmers have the telephone number of a local beekeeper contact and will inform the beekeeper when the application of insecticide is to take place. The beekeepers shut their hives during the days when spraying occurs. Spraying in the early morning or late in the evening also reduces the risk to bees and some insecticides are less harmful than others..

Insect growth regulators have been used with some success as they are highly specific and do not affect other groups of animals such as vertebrates, although they do affect useful insects as well as the pests. Compounds such as diflubenzron and methoprene mimic the action of the insect hormones affecting the development of the cuticle in the larval and adult stages in the life cycle. They can either prevent the larval stage developing into an adult by preventing the formation of an adult cuticle or prevent a new larval cuticle from developing after ecdysis.

Any chemical substance used to control pests affects the environment in some way and it is difficult to ensure that only the pest organism is affected, leaving the crop and soil uncontaminated. Many of the chemicals used have to be sprayed on to the crop, with the risk of contaminating plants and animals in the adjacent area. To ensure that there is enough of the chemical to kill or control the pest, there is a danger that excess may be applied and that some will get into water supplies and from there into food chains, with the added problems of bio-accumulation. Long-term use of an insecticide may result in insects becoming resistant and there is the danger of increased concentrations of the chemical being used in order to be effective. All chemicals used in the control of pests are now screened very carefully before being licensed by the EU or other governments. The chemicals have to meet strict standards of toxicity and speed of degradation Many of the chemicals used in the 1960s, 1970s and 1980s are now banned as they do not meet these standards.

Pesticide resistance will have a genetic basis, but it should be pointed out that the alleles conferring resistance are present in the pest population before it is exposed to the pesticide. The pesticide does not cause the mutations. When pesticide is applied to a crop, those members of the pest population that possess the alleles for resistance survive, reproduce and pass on their resistant alleles to their offspring, while the rest of the population perish. The frequency of the resistant alleles in the population increases unless the pesticide is changed or a combination of several chemicals is used.

As we have become more aware of the dangers of using chemicals, such as insecticides, fungicides and herbicides, polluting the environment and our food, alternative methods of protecting crop plants from damage and disease have been investigated. It is possible to reduce the effects of some pests by effective farming practices, such as good tillage to bury insect larvae and kill weed species. Suitable rotation of crops reduces or controls the numbers of pests and the planting of crops can be timed to avoid certain stages of a pest's life cycle. For such methods to be effective, a sound knowledge of the crop and the habits and life cycle of the pest organism are required.

Biological control involves a knowledge of the natural predators and parasites of the pest organism, which if properly applied can control or eliminate the pest without the need for chemicals. However, we need to be sure that the natural predator will only attack the pest organism and not create other problems in the environment. The predator should be able to grow well and reproduce in the same conditions that favour the pest, so that the control is maintained from season to season. There are many examples where natural predators have been introduced and have been successful in the control of a specific pest (ladybirds and hoverfly larvae eat aphids, *Cactoblastis* larvae feed on the prickly pear), but recent advances have focused on the use of microorganisms. Viruses and fungi have been used to control insect pests. A fungus, *Verticillium* sp., has been used to control aphids in glasshouses, the fungus producing enzymes which destroy the cuticle and exoskeleton of the insect, and sawfly infestations of spruce trees can be controlled by infecting the insects with a virus.

Biological control is not a quick solution to the pest problem, because the populations of predators take some time to build up and the damage to the crop might exceed the economic injury level before the predator is effective. It is also unsuitable if a crop is susceptible to a number of different pests. Eradication of a specific pest leaves its predator with no food and if the predator dies or moves away, the crop is left unprotected should re-infection occur later. Overall, the benefits outweigh the disadvantages and eventually such methods should prove less costly and more acceptable environmentally. Nowadays, biological control is more often used in combination with other methods, rather than as a method on its own.

Other methods of controlling insect pests, which involve a knowledge of their biology, include the sterilisation of males and the use of attractants, repellants, pheromones and insect hormones. Male insects can be sterilised by exposing them to X-rays, gamma rays or chemicals. They are then released into a population where they mate with normal females. This should result in a reduction in the population numbers, provided that sufficiently large numbers of sterile males can be reared and that the sterilisation procedure does not affect the mating behaviour of the males. It has been shown to work well for insects where mating takes place only once and has been successful in reducing screw worm in cattle in the southern USA.

Repellants and attractants have proved successful in specific cases. The broadcasting of bat calls in orchards has been shown to drive moths away and the use of tarry discs around cabbage plants prevents cabbage root flies from laying their eggs. Many insects can detect very small quantities of chemicals in their environment and control methods which make use of this to attract large numbers of insects to one place have proved very effective. Extracts from female insects have been used to attract males to a location where they are killed by a pesticide or sterilised. Some of the chemical substances used in this way are naturally-occurring pheromones, which the insects produce and which control their behaviour. As these pheromones are highly species specific and effective in small quantities, they are ideal for use in the control of insect pest populations. Pheromone attractants have been used to attract insect pests to monitoring traps. When the monitoring reveals a certain population level of the pest, the threshold level, it is economically worthwhile to spray with an insecticide. This method is particularly useful for the control of codling moth in apples and pea moth in peas.

No single method of control is 100 per cent effective, so the most appropriate approach is to combine suitable techniques and control measures in order to keep the pest population at a low level. The pest may not be eradicated, but its numbers are kept below the economic damage threshold. This form of control is termed **integrated pest management (IPM)** and involves a good understanding of the biology of the crop and the pest, together with a knowledge of the physical factors of the local environment.

An example of the use of integrated pest management is shown by the control of **whitefly** (*Trialeurodes*) infestations of glasshouse crops, such as tomatoes and cucumbers. The whitefly produces large numbers of flat, oval-shaped 'nymphs', which feed on the crop plants. A minute parasitic wasp (*Encarsia*) was found to parasitise the nymphs of the whitefly. The use of the parasitic wasp as a means of the biological control of the whitefly was investigated. It was observed that both wasp and whitefly numbers were affected by the environmental temperature. The whitefly grew and reproduced more rapidly than the wasp at 18 °C, but the wasp grew faster than the whitefly at 22 °C. At temperatures below 13 °C, the wasp will die. The density of the whitefly was also important. If wasps are to be introduced, the numbers of whitefly should not be too great as the wasps will be unable to control a sudden increase in whitefly population density.

A system of management is carried out where the crop is planted in treated potting compost in the glasshouse. The parasitic wasps are introduced, together with some whitefly to provide food initially. Provided that the temperature does not drop too low, killing the wasps, the whitefly can be controlled. If the numbers of whitefly do increase rapidly, or if the wasps are killed, insecticide sprays may be used. It is recommended that the insecticide used should be non-persistent, so that it will break down quickly, enabling rapid re-introduction of the wasp. This approach to the control of an insect pest makes use of a natural predator, takes into account the environmental factors and can be backed up by the use of an insecticide if necessary. In this way, less insecticide is needed, reducing possible residues on the crop, and the insect pest is less likely to develop resistance. This type of control also overcomes some of the drawbacks of using biological control on its own.

Integrated pest management was originally aimed at controlling individual pests of crop plants, but the more modern approach is to tackle all the pests that might affect a particular crop, becoming crop-specific rather than pest-specific. Much information about the conditions for growth and the pests of particular crops has been gathered and it is now possible to produce computer models of crop growth and the effects of environmental factors and pest development. Such models are useful for predicting the outbreaks of disease and could be developed further to include strategies for prevention and control.

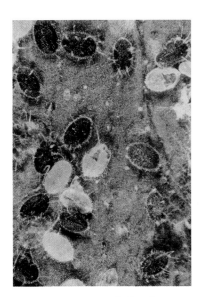

Figure 2.11 Parasitised whitefly nymphs (black) alongside non-parasitised nymphs (white)

Propagation and improvement of crop plants

Reproduction in flowering plants may be sexual or asexual. The flowering plant life cycle shows an **alternation of generations**, in which there is a spore-producing stage, called the **sporophyte generation**, alternating with a gamete-producing stage, the **gametophyte generation**. In flowering plants, the sporophyte generation is the dominant generation, producing the sexual reproductive stages in structures called flowers. Details of the flowering plant life cycle and the process of sexual reproduction are given in Chapter 6 of *Systems and their Maintenance*.

Green plants may also undergo a form of asexual reproduction, which is known as **vegetative propagation**. It involves the formation of specialised structures, sometimes referred to as **propagules**, derived from the root system or the shoot system. These structures can become detached from the parent plant and give rise to independent plants. In some cases, the development of the propagule involves the accumulation of a food store, resulting from photosynthesis of the parent plant. This food store enables the rapid growth of the propagule, following a period of unfavourable conditions, such as drought or low temperatures. Such propagules,e.g. bulbs of onions and strawberry runners, are typical of many herbaceous perennial plants and are referred to as **perennating organs**.

In sexual reproduction, meiosis occurs in the life cycle prior to gamete formation, so that each gamete has the haploid number of chromosomes. On fertilisation, a zygote is formed, the diploid number of chromosomes is restored and the resulting offspring are likely to show genetic variation due to the events of meiosis and the random fusion of the gametes at fertilisation. In vegetative (asexual) propagation, no gamete formation occurs, new individuals are formed from mitotic divisions and the offspring inherit identical genetic information from the parent.

Asexual reproduction

Asexual reproduction, in the form of vegetative propagation, is of significance in the production of some crops. The use of propagules has a number of advantages over sowing seeds, particularly where varieties have been developed for their high yield and resistance to disease. Because the propagules have the same genetic constitution as the parent from which they are derived, the grower knows that the crop will have the same desired characteristics. In addition, some propagules have a food store or already possess roots, so the crop becomes established more quickly than if seed is sown.

Strawberry plants produce modified stems, called **runners**, from axillary buds. These runners, which are **rhizomes**, grow horizontally over the surface of the soil, radiating out from the parent plant. Each runner bears axillary buds, which can develop adventitious roots and leaves, forming daughter plants. No food store is associated with this method of vegetative propagation as the daughter plants can obtain food from the parent plant until they have developed roots and leaves of their own. Once the daughter plants become established, the runners decay. The runners enable the new plants to become established at some distance from the parent plant, thus reducing intra-specific competition for light, water and nutrients. Each parent plant has the capacity to produce a large number of daughter plants vegetatively and strawberry growers make use of this method of propagation to clone large numbers of identical plants. Parent plants are selected for such favourable characteristics as high fruit yield, good flavour and resistance to disease.

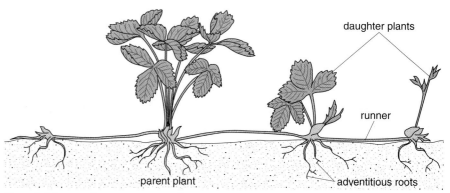

Figure 3.1 Runner formation in strawberries

Potatoes are grown for their stem tubers, which contain large amounts of starch (see Figure 1.6a). They are an important food crop as this starch is a cheap and easily cultivated source of carbohydrate in many human diets. The potato plant produces thin, underground stems, also called runners or rhizomes, which accumulate starch at the tips and swell up to form stem tubers. The aerial parts of the plant carry out photosynthesis and the soluble sugars are transported to the parenchyma tissue in the tuber, where they are converted into starch for storage.

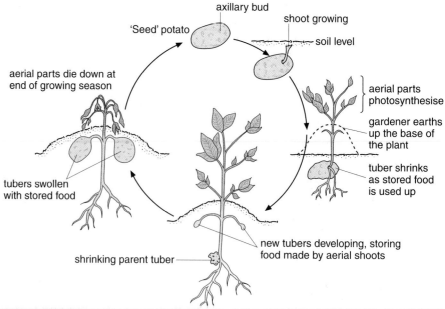

Figure 3.2 Tuber formation in potatoes

The potato grower plants small tubers in the spring, referred to as 'seed' potatoes. Aerial shoots and adventitious roots grow from axillary buds, called 'eyes', on the tubers, using the stored starch. When the new shoots are established and begin to photosynthesise, new tuber formation occurs. By planting 'seed' potatoes of a known variety from a reliable source, the grower is assured of the characteristics and quality of his crop.

Tillers are shoots which grow from axillary buds at the base of grass and cereal stems. As these shoots grow, adventitious roots develop at the base, anchoring the tillers in the soil. Each tiller could develop into an independent plant, but the tillers usually remain attached to the parent plant. This form of vegetative propagation is of some advantage to the cereal grower as each tiller is capable of producing a flowering spike, which after pollination and fertilisation, could give rise to an ear of the cereal crop. Reference has been made to tillers in Chapter 2.

In its dormant state, a bulb consists of a short conical stem bearing an apical bud, surrounded by swollen leaf bases containing a food store. There may be one or more lateral buds. When a bulb is planted, adventitious roots develop at the base of the conical stem and the buds grow, using the food stored in the swollen leaf bases. Leaves and a flowering shoot are produced by the apical bud. After flowering and seed dispersal, the foliage leaves continue to photosynthesise and the food is stored in the bases of these leaves, which swell up and become fleshy. A new apical bud develops and any lateral buds present become bigger and may form daughter bulbs, This sequence of events is shown in Figure 3.3.

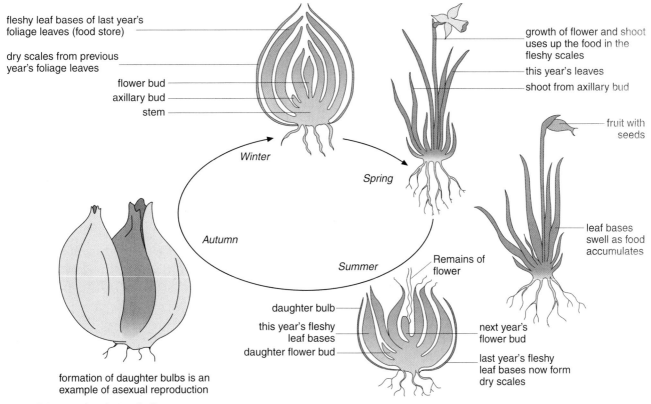

Figure 3.3 Annual cycle in a bulb

PROPAGATION AND IMPROVEMENT OF CROP PLANTS

Spring flowers, such as daffodils, narcissi and tulips are vegetatively propagated in this way. Onions are bulbs, but are grown commercially from seed. They are harvested at the end of the first growing season; humans making use of the food stored in the swollen leaf bases.

Underground rhizomes of plants, such as iris, are both over-wintering organs and organs of vegetative propagation. The swollen underground stems store food materials, enabling the development of axillary buds in the following growing season. Such rhizomes are of commercial importance in flower production.

In addition to natural vegetative propagation by such structures as bulbs, corms and tubers, other parts of plants will grow if detached and given suitable conditions. Some plants can be propagated readily from stem, leaf or root cuttings and whole plants may be regenerated from pieces of tissue or individual cells. This ability relies on cells being **totipotent**, each containing all the genetic information required to produce the new plant and also capable of division to form new cells. Such cells are found in **meristematic** tissues, in buds and in the vascular cambium, and are involved in the **artificial propagation** of crop plants by means of cuttings, grafting and micropropagation. There are several advantages to the crop producer in the artificial propagation of plants:

- Sexual reproduction introduces variation, whereas artificial propagation, involving only mitotic divisions, does not. For many crops, where uniformity is important, clones of identical plants are more easily handled than the offspring resulting from a sexual process.
- The juvenile stages of growth, involving only vegetative growth, are shortened. This has obvious advantages in growing fruit crops, where trees may grow for several years before producing flowers and fruit.
- The production of new plants by hybridisation may mean the fruits produced are sterile, incapable of producing viable seed. This feature has often been deliberately selected for in the case of bananas and seedless grapes, in order that the crop is more marketable, but it does mean that propagation has to be vegetative.

Cuttings

Cuttings may be taken from any part of a plant, and, provided that some vascular cambium or parenchyma tissue is present, given suitable conditions, they grow into complete new plants. The plants, or plant parts chosen may be in the juvenile phase of growth (this season's growth) or in the more mature phase of growth (last season's growth). With any type of cutting, rapid root growth is important, so that the developing plants can take up water and necessary mineral ions.

Soft tip cuttings are taken from young (juvenile) shoots which have developed little woody tissue. The cuttings are taken when the shoots are growing fast, usually in the spring, so that root development is rapid. The stem is cut just below a node, any leaves present are removed from the lower third of the cutting, which is then carefully placed into a prepared hole in potting

Humans utilise many parts of plants for food. Make a list of the more common vegetables, describe which part of each plant is eaten and evaluate its contribution to the diet.

compost. Often the cut end is dipped in fungicide, to prevent infection, but it is usually not necessary to use rooting hormone to stimulate root production. The cuttings are watered in and kept in conditions of high humidity to prevent the loss of water vapour from the leaves. Misting units, producing fine sprays of water, are used in the commercial glasshouse production of plants in this way, but enclosing a pot of cuttings in a polythene bag is equally effective. Soft tip cuttings are used to produce large numbers of chrysanthemum plants for the commercial market. These can be grown at any season of the year and induced to flower when required by manipulating the photoperiod.

Hardwood cuttings are taken in the autumn from mature plant stems, usually showing one year's growth,. The cuttings, about 20 cm in length, are cut from the base of the stem, just below a node. Any leaves present are removed from the lower parts of the cutting, the end of which is then dipped into rooting hormone, before planting out in compost or soil. The cuttings are usually planted so that only two or three buds show above the soil. No special precautions are taken to prevent loss of water vapour as these cuttings do not usually have leaves attached to them. When taken in the autumn or winter, the cuttings develop roots and new shoots during the following growing season. This method of propagation may be used for the rapid production of fruit bushes, especially blackcurrants. The new bushes are capable of producing fruit after one further year of growth, considerably more quickly than would be the case had they been grown from seed.

Grafting and budding

The technique of **grafting** is used to join a shoot, known as the **scion**, from one plant to the root system of another plant, known as the **rootstock**. It is a method commonly used to propagate fruit trees, when stem cuttings are not successful. The scion is taken from a selected variety, known as a **cultivar**, and the end is cut so that it fits closely into a specially-shaped groove on the rootstock. The cuts expose the vascular cambium of both the scion and the rootstock and for the formation of a firm union it is essential that there is good contact between the two. The join is bound tightly and made waterproof, so that water loss is reduced and infection by pathogenic microorganisms is avoided. Initially the cambial cells produce callus tissue, which unites the scion and the rootstock, then new cambium develops, capable of producing new xylem and phloem tissue and completing the link between the two parts.

The technique takes time to carry out and can be difficult, but it enables the selection of the scion for desirable characteristics, such as fruit type and quality, yield and disease resistance, as well as a choice of rootstock, which can determine the size of the mature tree. For example, dwarfing rootstocks are popular in small gardens, where space is limited, and they are also favoured by apple growers where ease of harvesting is important. It is possible to graft several varieties of apple on to one rootstock, so that fruit suitable for eating and cooking can be grown on the same tree. As apples are self-sterile, it is useful to graft pollinator varieties on to the same tree.

Figure 3.4 Chrysanthemums commercially produced by soft-tip cutting.

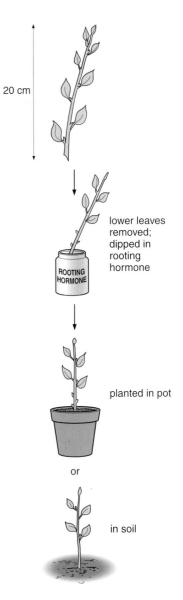

Figure 3.5 Taking hardwood cuttings

20 cm

lower leaves removed; dipped in rooting hormone

ROOTING HORMONE

planted in pot

or

in soil

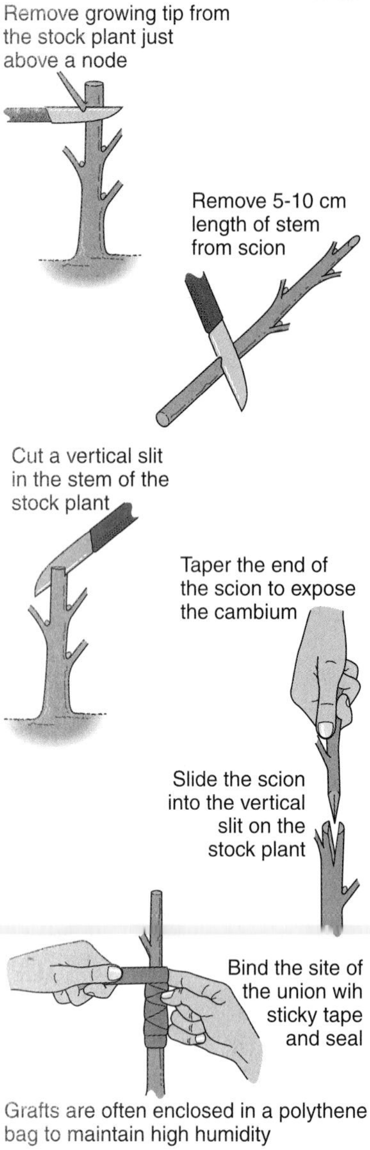

Remove growing tip from the stock plant just above a node

Remove 5-10 cm length of stem from scion

Cut a vertical slit in the stem of the stock plant

Taper the end of the scion to expose the cambium

Slide the scion into the vertical slit on the stock plant

Bind the site of the union wih sticky tape and seal

Grafts are often enclosed in a polythene bag to maintain high humidity

*Figure 3.6 **Making a graft***

Budding is a technique, similar to grafting, used in the propagation of roses, which do not breed true if grown from seed. Vegetative buds are carefully removed from the desired variety and inserted into special cuts made in the bark of the rootstock. This is a rapid means of producing large numbers of plants when demand is high.

Micropropagation

Micropropagation involves the use of **tissue culture** for the rapid production of large numbers of genetically identical plants of the same variety. Cambium tissue from meristematic regions of plants is cultured on media containing nutrients and growth substances under carefully controlled, sterile conditions. In commercial crop propagation, sections of stem with an apical meristem, called **explants**, are used, but in some cases it is possible to use small sections of leaf. The tissue is surface-sterilised, using dilute hypochlorite solution, rinsed in sterile water and then placed on a sterile culture medium. These operations are carried out in aseptic conditions, often in a laminar flow cabinet, where the incoming air is filtered to remove spores, which might contaminate the cultures. Sterile instruments are used when handling the tissue and protective clothing is worn. The culture medium used contains a high concentration of sucrose and a range of minerals and vitamins incorporated into agar, so that the developing shoots are supported and do not sink into the medium where conditions could become anaerobic. Plant growth substances are used to promote root and shoot development. The explants are kept in sealed containers in a growth room where the temperature and light conditions are controlled (Figure 3.8).

Initially, a mass of undifferentiated cells is produced, which develops into a group of tiny plantlets. At this stage, the level of cytokinin in the growth medium is high, promoting the formation and development of axillary buds, so that a short shoot with many branches is produced. At intervals of about a month, the cultures are transferred to fresh growth medium. During this procedure, the cultures are divided by separating out the branches, thus increasing the numbers of plants. When a large number of plants has been produced in this way, they are transferred to culture media containing a high auxin to cytokinin ratio in order to promote the formation of roots. With a reduction in the concentration of cytokinin, the shoots grow taller and axillary buds remain dormant.

When the shoots and roots are sufficiently developed, the plants are removed from their containers and planted out in a glasshouse. They have to be kept in a humid atmosphere, with controlled lighting, until the waxy cuticle develops and they begin to photosynthesise.

Sexual reproduction

Sexual reproduction in flowering plants involves the formation of gametes, pollination and fertilisation, followed by fruit and seed production. Details of the flowering plant life cycle and flower structure are given in Chapter 6 of *Systems and their Maintenance*.

*Figure 3.7 **Plantlets produced by tissue culture.***

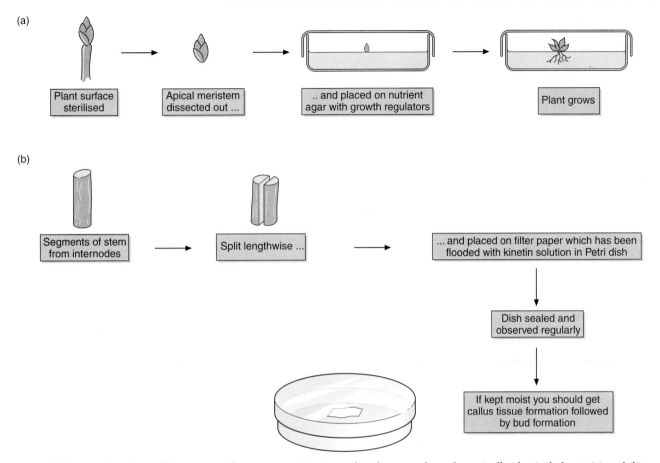

(a)

Plant surface sterilised → Apical meristem dissected out ... → .. and placed on nutrient agar with growth regulators → Plant grows

(b)

Segments of stem from internodes → Split lengthwise ... → ... and placed on filter paper which has been flooded with kinetin solution in Petri dish

Dish sealed and observed regularly

If kept moist you should get callus tissue formation followed by bud formation

Figure 3.8 Tissue culture is used in commercial crop propagation to produce large numbers of genetically identical plants; (a) and (b) above show two different ways of carring out micropropagation.

The microspores, or pollen, contain the male gametes and develop in the pollen sacs of the anthers of a flower. An anther, supported by a stalk called the filament, usually contains four pollen sacs. In the early stages of their development, each pollen sac contains a central mass of diploid microspore mother cells, surrounded by a nutritive layer called the tapetum. The microspore mother cells undergo meiosis, producing tetrads of haploid cells, each of which undergoes further development into a microspore, or pollen grain. Within each pollen grain, the haploid nucleus divides mitotically to produce a generative nucleus and a pollen tube nucleus. The generative nucleus undergoes a further mitotic division, producing two nuclei which function as male gametes. Each mature pollen grain is surrounded by a thin, inner wall, called the intine, and a thick, outer wall, the exine. In wind-pollinated species, the exine is smooth, but in insect-pollinated species the exine is often pitted or sculptured.

As pollen formation proceeds, the cells in the tapetal layer shrink and break down, and fibrous layers develop in the anther wall. As the anther dries out, tensions are set up in the wall, which eventually splits. The two edges curl away from each other, exposing the mature pollen grains.

When mature pollen grains are ready for dispersal, they have a water content of between 10 and 15 per cent, which is similar to that of seeds. They also

PROPAGATION AND IMPROVEMENT OF CROP PLANTS

The ovules, containing the female gamete, or egg cell, develop in the ovary of the flower. Each ovule begins as a tiny structure, called the nucellus, in which there is a megaspore mother cell. Following the meiotic division of this megaspore mother cell, one of the haploid megaspores produced undergoes further divisions to form an embryo sac, containing eight nuclei, each surrounded by a small quantity of cytoplasm. At one end of the embryo sac, the female gamete, or egg cell, is flanked by two cells called the synergids. At the opposite end is another group of three cells, called the antipodal cells. Two nuclei, the polar nuclei, remain in the centre. These nuclei may fuse to form a central diploid (fusion) nucleus. Prior to pollination and fertilisation, a mature ovule consists of the nucellus, containing the embryo sac surrounded by two protective layers of cells, the integuments. These layers do not completely enclose the ovule, as there is a small opening, called the micropyle, situated near the embryo sac.

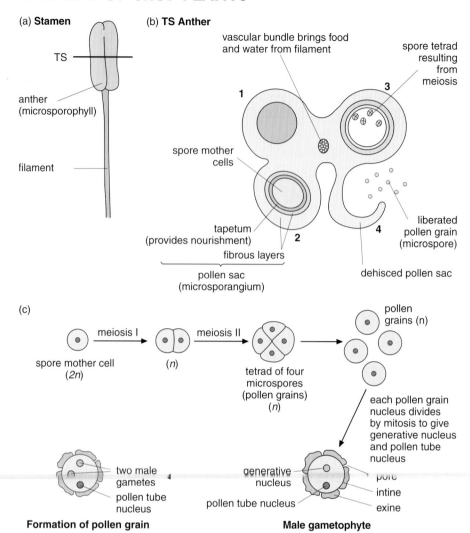

Figure 3.9 Pollen formation and structure

have stores of carbohydrate, lipid and protein, which are mobilised on germination and provide the necessary energy for the growth of the pollen tube until it obtains nutrition from the tissues of the style. At this stage the pollen can be considered to be dormant, but viable.

Pollination is the process by which pollen grains, containing the male gametes, are transferred from the ripe anthers to the receptive stigma. In some species, **self-pollination** occurs, where pollen grains are transferred to the stigma of the same flower, or to another flower on the same plant. In other species, pollen grains from the anthers of one flower are transferred to the stigma of a flower on another plant of the same species. This process is known as **cross-pollination**.

The transfer of the pollen grains is achieved by air movements in the case of wind-pollinated flowers or by insects. The characteristics of wind- and insect-pollinated flowers are described fully in Chapter 6 of *Systems and their Maintenance*.

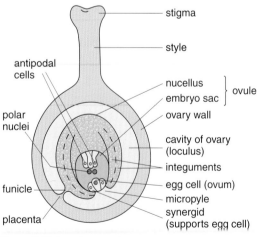

Figure 3.10 A mature ovule

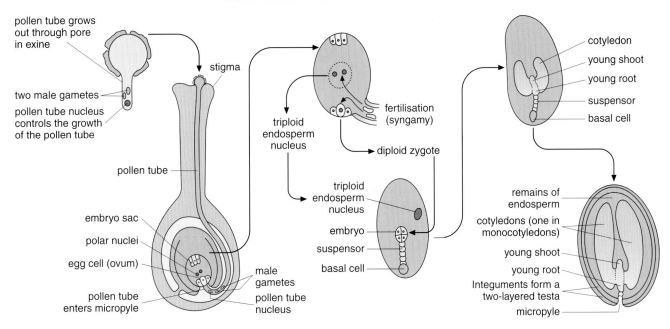

Figure 3.11 Fertilisation

The epidermal cells of the stigma secrete a solution of sucrose, which stimulates the germination of mature pollen grains. Each pollen grain is able to produce a pollen tube, which grows out through a pore in the wall. The growth of the pollen tube is controlled by the tube nucleus. Digestive enzymes, which soften the cutin of the stigma and the middle lamellae of the cell walls, are secreted and enable the pollen tube to penetrate the tissues of the style and grow down towards the ovary. The pollen tube grows towards the ovules, attracted by a substance produced by the micropyle.

The **viability** of pollen can usually be determined by providing conditions suitable for its germination. Viable pollen should germinate in dilute sucrose solutions. Methods, such as the use of tetrazolium salts or catalase activity, similar to those used to determine seed viability, can prove difficult due to the small size of the grains. A method, based on a fluorochromatic reaction catalysed by enzymes in viable pollen grains, was devised by Heslop-Harrison in the 1970s. As it requires a fluorescence microscope to determine the results, it is impractical for use in schools.

Fertilisation is achieved when the tip of the pollen tube enters the ovule through the micropyle and comes into contact with the embryo sac. The male nuclei are released through a pore which develops in the tip of the pollen tube. One of the male nuclei fuses with the egg cell, forming a diploid zygote, and the other male nucleus fuses with the two polar nuclei, or the central diploid (fusion) nucleus, to form a triploid primary endosperm nucleus. The endosperm nucleus gives rise to the nutritive tissue, known as the endosperm, during the development of the seed.

Self-pollination leads to self-fertilisation and the offspring of plant species which are self-pollinated usually show little variation due to the recombination of genes from a single parent. They are described as **inbreeders**. If an individual plant, heterozygous at a single locus (**Aa**), is repeatedly self-pollinated, it will become homozygous at that locus and two pure-breeding lines **AA** and **aa** will eventually become established. Inbreeding results in uniform populations, the only variation resulting from chance mutations. This can be an advantage to the producer, where desirable characteristics have been selected for in crops such as cereals and some legumes, which are self-pollinated and hence inbreeders.

Many plant species have evolved mechanisms which prevent self-pollination, so that fertilisation following cross-pollination gives rise to **outbreeding** and offspring which show variation. Outbreeding maintains heterozygosity, because two different parent plants, with different genotypes, are involved. For details of the mechanisms which prevent self-pollination, reference should be made to Chapter 6 of *Systems and their Maintenance*. Both inbreeding and outbreeding play significant roles in the development of improved crops.

Seed formation and embryo development

After fertilisation has occurred, the fertilised ovule develops into a seed and the ovary develops into a fruit. Within the ovule, the diploid zygote undergoes mitotic divisions and a multicellular embryo is produced. The embryo consists of:

- a **plumule**, which becomes the first shoot: it consists of a stem with a terminal bud surrounded by the first pair of foliage leaves
- a **radicle**, which becomes the first root
- one or two **cotyledons**, known as seed leaves: these are simpler than foliage leaves and may contain food stores for the embryo during germination.

The embryos of monocotyledonous plants have a single cotyledon and those of dicotyledonous plants have two. Very young plants can be recognised by their cotyledons alone, as the shape of the cotyledon is specific to each species. This characteristic is valuable to the crop producer as it enables the early recognition of weed species amongst the crop. The earlier the weeds are recognised, the easier it is to control them.

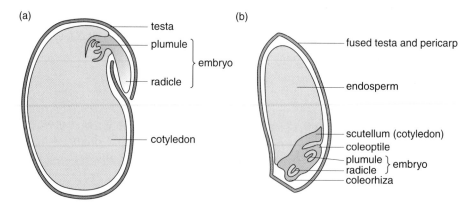

Figure 3.12 (a) legume seed and (b) cereal grain

Within the endosperm, mitotic divisions give rise to large numbers of triploid nuclei. Each becomes separated from the others by thin cell walls. In some seeds, such as cereal grains, this tissue forms the food store, but in others, such as peas and beans, food is stored in the cotyledons, which swell up and become fleshy. The food stores are produced as a result of photosynthesis in the parent plant. The soluble products of photosynthesis are translocated in the phloem to the developing seeds, where they are converted into insoluble storage compounds, such as starch. As the embryo grows and the food store develops, the seed gets bigger. The integuments surrounding the ovule become the **testa**, or seed coat, which forms a tough protective covering for the seed during its period of dormancy. The micropyle persists as a tiny pore in the testa, through which water is taken up when germination occurs. The water content of the seed is reduced in the later stages of its development to between 10 and 15 per cent of its mass.

In leguminous crops, which are members of the Papilionaceae, the seeds are contained in pods, which become dry and split when mature, releasing the seeds. In cereal crops, which are members of the Gramineae, the fruits are one-seeded and the pericarp is fused with the testa of the seed forming the outer coat of the grain. In wild varieties of cereals, the central stalk of the flowering spike breaks up, releasing the grains, but the cultivated varieties have been selected so that this stalk does not shatter and the ears of grain can be harvested intact.

Seed germination

Some seeds can germinate immediately after they have been shed from the parent plant, but most seeds require a period of dormancy before germination will occur. Some need to be kept dry or require exposure to a period of low temperature. Light conditions, temperature and growth inhibitors may be involved in seed dormancy. It is of interest to note that the seeds of many plants may germinate successfully, but the resulting plants need an environmental signal, such as cold treatment or changes in day length, before they can produce flowers. Winter varieties of barley, rye and wheat require a period of low temperature to cause production of flowering shoots. If these varieties are planted too late in the spring or in tropical climates, then their growth remains vegetative. Spring varieties of these cereals do not have this requirement, but do not give such high yields of grain. Biennial crops, such as sugar beet, must not be exposed to frost after emergence. If a late frost occurs, such plants may enter the flowering phase in the first year instead of producing the normal swollen root, a process known as bolting.

Viable seeds require water, a suitable temperature and oxygen in order for germination to occur. Water is taken up through the micropyle in a process called imbibition. This causes the seed to swell and the testa ruptures, allowing first the radicle and then the plumule to emerge. Water activates the hydrolytic enzymes, which results in the breakdown of the food reserves in the seed, providing energy and materials for the growth of the embryo into a seedling. More details of the process of germination are given in Chapter 6 of *Systems and their Maintenance*.

For many crops, the producers require a supply of good seed, which will grow into plants which are true to type. Named varieties of crops are often chosen to suit the environment in which the crop is to be grown. The producer needs to obtain seed from a reliable source as it is unusual nowadays to save seed from a previous crop. Certified seed, produced by specialists under strictly controlled conditions, is available and this seed should be:

- free from weed seeds
- free from the seeds of other varieties
- clean
- disease- and pest-free.

Certified seed is guaranteed to produce a crop true to type and give a high germination rate.

Seed viability, percentage germination and **seed vigour** are of importance to the crop producer. If a large number of non-viable seeds is planted, then the percentage germination is low and crop yield will eventually be affected. It is possible to apply simple tests for seed viability, using tetrazolium salts to stain living tissue or by determining catalase activity. The results of these tests indicate the presence or absence of metabolic activity in the tissues of the embryo, but this does not necessarily mean that all the seeds will germinate. Percentage germination can easily be determined by giving seeds the correct conditions and then counting the number which have produced radicles in a given time. Such determinations of percentage germination are usually carried out under laboratory conditions, so give no indication of the ability of the germinated seeds to grow into plants in the field. An indication of seed vigour is given by the speed at which germination occurs and this is affected by the size of the seed, its age and whether it is infected with pathogenic organisms.

Improvement of crop plants

Farmers have been selecting suitable varieties of crops for a very long time. Seeds from plants with desirable characteristics were saved after harvest and planted the following year. When the plants grew and produced flowers, they were cross-pollinated with others showing the same characteristics. After many generations of this process of cross-pollination and selection being repeated, whole populations would show the desirable characteristics, producing pure lines. The development of pure lines in this way results in the plants tending to become homozygous for less desirable characteristics as well as for the desirable ones. This strategy could eventually lead to **inbreeding depression**, which could result in a reduction in the crop yield due to low seed germination rates, lack of resistance to disease or less vigorous plants.

In some breeding programmes, two inbreeding varieties with desired characteristics are crossed to produce **hybrids**. Self-pollination has to be prevented and this can be achieved by **emasculation**, which involves the removal of the male reproductive structures, the stamens, before the pollen is mature. This procedure must be carried out carefully, using a needle and forceps, in hermaphrodite flowers, where male and female reproductive structures are close together. Where there are male and female flowers, as in maize, it is easy to remove the male flowers from the plants. When the female reproductive structures are mature, pollen grains from the ripe anthers of the chosen parent are transferred to the ripe stigmas of the emasculated flowers, using a small paint brush. Care has to be taken to ensure that the stigmas are not contaminated by pollen from any other source. Immediately after pollination, the flowers are enclosed in pollen-proof bags. Reciprocal crosses, reversing the roles of 'male' and 'female' parents, are usually carried out.

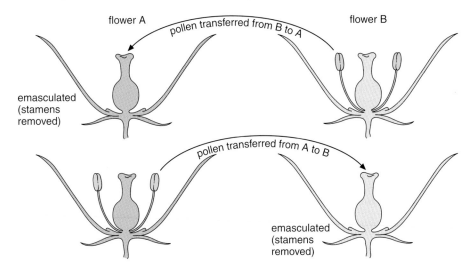

Figure 3.13 Controlled pollination

The seeds produced, referred to as **F$_1$ hybrids**, are collected when ripe. These seeds are usually more vigorous than the normal seed. When this seed is sown, the plants show **hybrid vigour**, or **heterosis**, which could result in increased yields for the crop producer. The inbred parents are homozygous at many gene loci and the offspring of the cross, the F$_1$, will show genetic uniformity. If self-pollination of the F$_1$ occurs, there is increased variability in the offspring as the gene loci tend to be heterozygous. This means that the seed from F$_1$ hybrids cannot be saved to produce a crop with the same characteristics the following year: new seed must be purchased.

Polyploidy

Plants with three or more complete sets of chromosomes are known as **polyploids**. Many important crop plants are polyploids and the condition is associated with increase in size, increased hardiness and resistance to disease. There are two types of polyploidy: **autopolyploidy** and **allopolyploidy**.

Autopolyploidy involves an increase in the number of sets of chromosomes in the same species and may arise naturally when a cell undergoes abnormal meiosis. If the chromosomes fail to separate during prophase I, this results in a gamete having the diploid number (2n) of chromosomes rather than the haploid number (n). If this gamete fuses with a normal haploid gamete, the zygote would possess three complete sets of chromosomes and become triploid (3n). It is possible to induce abnormal meiosis, using a substance called colchicine, which interferes with the formation of the spindle in dividing cells. As a result, chromosomes do not separate and move to opposite poles

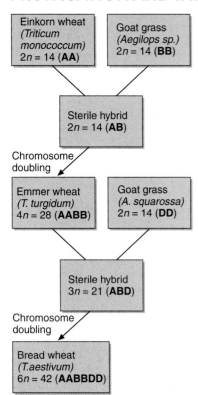

Figure 3.14 Development of hexaploid wheat

and tetraploids may be formed. This technique has been used by plant breeders to produce improved varieties of tomatoes and sugar beet.

In allopolyploidy, the chromosome sets come from different species. When two different species are crossed, the resulting F_1 hybrids are usually sterile because the chromosomes cannot pair up during prophase I of meiosis. If the chromosome number is doubled, each chromosome now has an homologous partner and behaves in the same way as a diploid, so that fertile hybrids can be produced. Plant breeders can use this technique to combine different characteristics from different, closely related species. An example of a successful application of this technique is the development of the cereal crop *Triticosecale* (*Triticale*), which combines the yield and grain quality of wheat (*Triticum*) with the hardiness of rye (*Secale*).

Modern wheat, which is hexaploid (6n), is thought to have evolved by hybridisation and chromosome doubling. Different varieties of diploid wheats were known to have been cultivated by humans about 10 000 years ago and it has been possible to trace the history of the development of modern wheat from diploids to tetraploids, and from tetraploids to hexaploids. A possible sequence of this process of hybridisation and doubling of chromosome numbers is shown in Figure 3.14.

Many crops, such as sugar cane, tobacco and oats, have evolved as allopolyploids and more recently other hybrids have been produced by plant breeders. Varieties of wheat with short stems and high-yielding varieties of rice have been developed to suit modern requirements. Sometimes the improvement may mean easier cultivation or harvesting rather than the emphasis being on bigger, better plants.

Seed banks

Modern farming practices, with their emphasis on high yields, result in a reduction in the varieties of different crops available to the consumer. This is clearly evident in the number of different varieties of apples for example, on display in supermarkets in the UK. For most crops, only a small number of varieties is now available to the producer. These have usually been bred for their high yield and resistance to specific diseases. This reduction in diversity could eventually be disadvantageous to plant breeders, so seed banks have been set up to preserve large numbers of different varieties. In this way, a reservoir of genes is maintained and plant breeders have the material available from which to select advantageous genes, retaining the potential for breeding new varieties. Seeds can be kept in their dehydrated state for a long time and it is also possible to preserve living tissue in cold storage. Care must be taken to grow these varieties from time to time, in order to preserve their viability.

Gene technology

The introduction of new genes into crop plants is of great interest to plant breeders. It allows the selection and direct introduction of desired characteristics into plants, without having to wait for the results of traditional techniques such as hybridisation and the formation of polyploids. The traditional techniques may take several growing seasons before the improved varieties are of benefit to the crop producer. The basic techniques of the isolation of genes and their incorporation into bacterial plasmids are described in Chapter 7 of *Cell Biology and Genetics*.

Much interest has been shown in the soil bacterium *Agrobacterium tumefaciens*, which causes crown gall disease in dicotyledonous plants. This bacterium gains entry to the plants through a wound and stimulates the production of a tumour, referred to as a gall, in the stem. The growth of the tumour tissue is due to the presence in the bacterium of a plasmid, the T_i plasmid. A piece of this plasmid can become incorporated into the DNA of the host plant cells, where it replicates and causes the host plant to release hormones, which stimulate the production of the cells forming the tumour. It has been found possible to isolate this plasmid and introduce desirable genes into it. The plasmid is then used as a vector to carry the genes into plant tissue. The resulting tumour tissue can be used to produce **transgenic plants** containing the desirable genes.

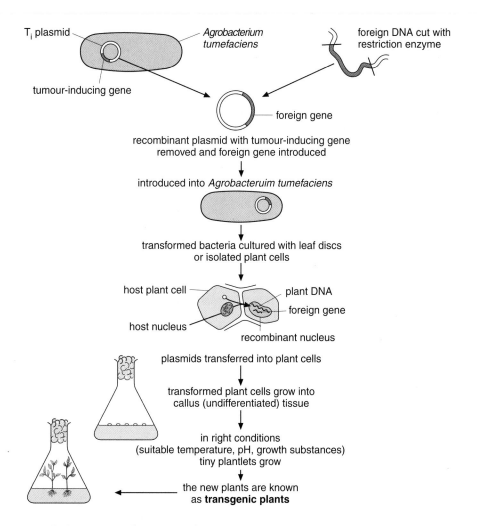

Figure 3.15 Formation of transgenic plants

PROPAGATION AND IMPROVEMENT OF CROP PLANTS

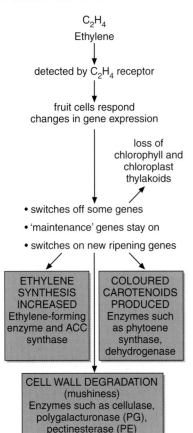

Figure 3.16 *The role of ethylene in the control of ripening in tomatoes*

A great deal of research is currently being carried out using tomatoes, potatoes, sugar beet and other plant crops, but the technique does depend on the identification and isolation of the desirable genes, which is a time-consuming process.

Transgenic tomatoes have been created by modifying the genes involved in the ripening processes, thus delaying the onset of spoilage associated with ripening yet retaining the colour, aroma and desirable flavour. Ethylene has an important role in switching the genes involved in ripening on and off. The sequence of events is shown in Figure 3.16.

Pectinesterase (PE) and polygalacturonase (PG) are both enzymes involved in the breakdown of pectic substances in the cell wall, which lead to 'mushiness' in overripe tomatoes. PG is absent in unripe fruit, or present only in very low quantities, but occurs in large quantities in ripe fruit. An 'antisense' PG gene has been inserted into strains of tomato, resulting in reduced PG production and delay in the development of 'mushiness'. The tomatoes stay firm while ripening and they can be kept longer on the plant before picking. The tomatoes have increased flavour, as they can be picked later when they are more mature, and they are less likely to be damaged in transit.

In the USA, soya beans have been genetically modified to be resistant to glyphosate, a systemic (translocated), non-selective herbicide. The crops can be sprayed with this herbicide, which will kill all other weeds. Glyphosate is an environmentally safe herbicide, as it is broken down to harmless compounds by microorganisms in the soil and has a low toxicity to mammals. There has been some concern about the use of this genetically modified soya and some firms in the UK and EU are refusing to buy it. In the USA it is often mixed with unmodified soya and so difficult to trace. Strawberries have been bred to be resistant to vine weevil and apples have been modified so that they stay firm longer in storage, but neither of these modifications are commercially available yet.

Use has been made of the ability of the soil bacterium *Bacillus thuringiensis* to produce proteins known as T-toxins, which kill insects, but are harmless to vertebrates. A gene for one T-toxin has been successfully cloned into maize and tomato plants, conferring protection against insect pests.

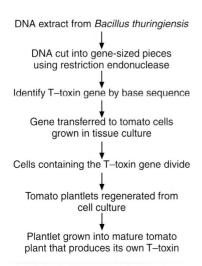

Figure 3.17 *Transfer of T-toxin into tomato plants*

Cattle and sheep

The cattle and sheep industry

Animal production systems provide a means of converting plant products into animal products which humans can exploit. In modern farming, the emphasis is on animal products as part of the 'human food chain', though from early times, domesticated cattle and sheep have been important to human communities for other reasons, including for power, for clothing and often they had a religious significance.

Origins

Cattle and sheep have been associated with human communities from early Neolithic times. They are ruminants and digest cellulose so do not compete directly for the same food resources as humans. They have a natural social instinct, enabling them to be domesticated successfully and this has persisted in their herd behaviour today. Sheep were among the earliest animals to be domesticated and their origins are believed to be in western Asia around 9000 years ago. Cattle were probably also first domesticated in western Asia, about 1000 years later. There is evidence of milk production from cattle in Egypt and Mesopotamia about 6000 years ago and weaving with wool at about the same time. The wild ancestor of sheep is believed to be the Asiatic mouflon (*Ovis orientalis*) and of cattle, the now extinct aurochs or wild ox (*Bos primigenius*).

When first domesticated, cattle were probably valued as power animals and for their religious significance rather than for food. Over the centuries, cattle have been exploited in various ways - as draught animals for drawing a plough or cart; as food (meat, milk, dairy products); clothing or leather from the hide; weapons or artifacts from the bones and horns; tallow for candles and soap from the fat - and gave a measure of wealth. The dung provides manure which has been important for maintaining soil fertility, for making bricks for building and as a fuel. With sheep, there has been progressive selection for fleeces that are white and grow continuously allowing them to be shorn. The wool derived from the fleece has allowed development of clothing fibres and textile industries that have had unique importance in the economy of many countries through the centuries, including ancient Mesopotamia, medieval Britain and modern Australia. Sheep remain important in many areas of the world for providing meat and milk (hence yoghurt and cheese), and wool and skins for clothing.

Development of breeds of cattle for meat and milk

During domestication of cattle, two distinct lines have emerged – the humped (zebu type) and the humpless, the latter being further subdivided into longhorned and shorthorned cattle. The humped form is distributed predominantly in Asia and parts of Africa, its main use being for cultivation with the plough and some for meat. The humpless form occurs throughout Europe, in northern Africa and northern Asia, regions with a much more developed dairy industry.

Figure 4.1 (top) Using humped cattle for ploughing in rice paddy, Yunnan province, south-west China; (bottom) a sheepskin over the back is a decorative part of clothing for Nakhi women in south-west China. It also protects the back when carrying heavy baskets.

CATTLE AND SHEEP

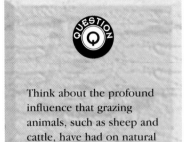

QUESTION

Think about the profound influence that grazing animals, such as sheep and cattle, have had on natural ecosystems.

In the UK, different factors have influenced the development of breeds leading to the structure now found in the modern cattle industry. Before the 18th century in England, cattle were used as working animals and to provide tallow, so selection was mainly for size and fatness. Then horses became more commonly used for work, and selection of cattle for production of meat and milk became a higher priority. Enclosure of common land to form discrete fields, allowed more control over the animals and adoption of better feeding practices encouraged more sophisticated breeding programmes. The performance of the breeding animals and of their offspring began to be carefully documented, the beginning of record systems used extensively today. Since that time, there has been a succession of breeds, reflecting changing uses, development of superior performance and generally greater specialisation.

In 18th century Britain, the Longhorn was the dominant breed but this is now a rare, nearly extinct breed. The Shorthorn took over from the Longhorn because it was useful for both milk and beef. The Shorthorn remained important well into the twentieth century. In 1908, Shorthorns accounted for 64% of cattle in Britain but this fell to 50% by 1936 and has now dwindled to 1%. This reduction is largely due to the introduction of black and white Friesians from Holland around the beginning of the twentieth century. Friesians became successful because they produce more milk and more beef than Shorthorns. Selection in the Holstein breed, in both North America and Europe, has been primarily for milk production, with some loss of yield and quality of its beef. Friesian and Holstein breeds and their crosses now account for the majority of cattle in Britain. Local breeds do still exist in Europe, but there is increasing concentration on just a few successful breeds.

Figure 4.2 (left) 19th century drawing of Longhorn cattle – once dominant in England, now a rare breed; (right) the Presba cow – a rare diminutive shorthorn breed, just over 1m in height, from north-west Greece.

The modern cattle industry shows some separation of production into dairy cattle and beef cattle. In continental Europe, breeds are often dual purpose, so the same cattle are used for milk and for beef production. In the USA, there is greater polarisation of breeds into those showing characteristics appropriate for milk or for beef. In the UK, about 60% of beef comes from the dairy herd, so only about one-third is raised purely for beef (though less since BSE).

QUESTION

Specialisation of breeds brings the potential danger of loss of genetic diversity. What are these 'dangers'? What measures can be taken to maintain genetic diversity?

Various factors are considered when selecting breeds to be used in a herd and the bulls and cows to be mated within the breed. In dairy cows, selection is made for milk yield (volume per day), length of lactation, quality of milk

Table 4.1 *Some features of breeds of cattle in the UK, including some recently introduced continental breeds*

Name of breed	Type	Colour	Features
Friesian	dual purpose	black and white	• originated in Holland (Friesian Islands) • high milk yield plus good lean meat • quick-growing large animal • highest proportion of UK herds
Holstein–Friesian	mainly dairy	black and white	• a strain of Friesian, imported from North America • very high milk yield • large bony frame, lean meat
Ayrshire	dairy	white and brown	• originated in Scotland • good quality milk, less suited to beef production • does well under poor conditions
Jersey	dairy	light fawn, to dark brown or grey	• originated in Channel Islands • high butterfat content in milk • milk used particulary for butter and cream
Shorthorn	dual purpose	red, white, roan	• formerly main breed in Britain, now numbers very low
Hereford	beef	red, with white face and chest	• chief beef breed in Britain, popular for crossing with other breeds • large, hardy grazing animal • early maturing
Aberdeen Angus	beef	black	• originated in Scotland, hornless • very high quality quality beef breed, needs good feeding • early maturing, high proportion meat to bone
Galloway	beef	black	• originated in Scotland, hornless • hardy grazing breed, suited to poor conditions
Highland	beef	brown	• originated in Scotland, very long horns, long shaggy coat • very slow growing • suited to poor conditions
Charolais	dairy	creamy white	• originated in France • large heavy cattle, with large hindquarters • fast growth, lean meat, low fat
Limousin	beef	red	• originated in France • very large hardy breed • fast growth, good quality meat carcass
Simmental	dual purpose	yellowish brown or red	• originated in Switzerland • good milk yield, good carcass when reared for beef • high growth rate

(butterfat the and protein content) and susceptibility to mastitis. Important physical factors include the shape of the udder and its support, and the length and angle of the teats to ensure ease of milking. Consideration must also be given to externally imposed factors, such as regulations governing market prices and milk quotas. A herd of cows may, for example, yield the permitted volume of milk but be penalised for a high fat content.

In beef cattle, selection is generally for a high yield of good quality meat, considered in relation to the feeding system used. A large calf is likely to reach a large body size with a high yield of meat, but the large size may cause difficulties at calving. Some breeds have a high calf mortality rate. Particularly important is the rate of growth of the calf and likely weight at slaughter at a certain age. During growth and development, deposition of fat occurs after bone and muscles have been laid down - if energy intake exceeds that required to lay down bone or muscle, fat is deposited. The age of maturity is linked to development of lean or fat meat. Late maturing breeds tend to have high growth rates with deposition of fat occuring at a later age and higher body weight compared with early maturing breeds fed on the same diet. To illustrate this, we can look at bulls used for beef production. In Britain, the Hereford and Aberdeen Angus have been used and both begin their fattening phase relatively early. This suited a fattening system based on grass, but with more intensive systems of feeding, involving cereal, Hereford and Aberdeen Angus develop too much fat at lighter weights. Increasing consumer demand for lean meat rather than fat, has led to a shift towards continental breeds, such as the Charolais and Limousin. These have become popular because they are larger, provide a high proportion of very lean meat and, even though late-maturing, they have a high muscle to bone ratio which is an advantage.

Table 4.2 *Some effects of the breed of different bulls - these are only some of the factors to be considered when choosing a breed for mating*

Breed of bull	Calf mass at 200 day / kg			Mass at slaughter / kg	Assisted calvings (%)	Calf mortality (%)
	Lowland	Upland	Hill			
Hereford	208	194	184	410	3.8	1.6
Angus	194	182	176	393	3.1	1.3
Charolais	240	227	205	494	9.6	4.8
Limousin	215	204	186	454	7.2	4.4

The effects of the breed of different bulls in relation to some of these factors are illustrated in Table 5.7

In choosing which individual bull and cow to be mated, or which bull semen to use by AI, the farmer needs to have some indication of likely performance of the offspring. Actual performance of a bull in a herd cannot be predicted precisely because of natural variation, the effects of environmental factors and systems used for management of the herd. However, useful information can be gained from

- **detailed records of performance** of parents and grandparents, though this does not necessarily mean that the characters are passed on to their calves
- **progeny testing** which monitors the performance of the bull's calves, compiled from a series of matings, on different farms, with the progeny being raised in a variety of environmental conditions and different systems of management. Progeny testing is inevitably slow because it takes time to rear the calf to adult when the outcome can be measured.

Information relating to the genetic characteristics of a very large number of animals can now be stored on computer databases, taking into account environmental conditions which would influence expression of particular traits. Farmers seeking genetic improvement of the herd, or wishing to manipulate the characteristics in a particular way, can make a reasonably informed choice of bull, or semen to inseminate the cows. Information about each bull is continually updated. Evidence of superior performance usually leads to a higher price for the semen offered for artificial insemination. Costs of desirable bull semen can be very high, so this must be considered with other factors in the management of the herd. A computerised scheme known as **BLUP** (**B**est **L**inear **U**nbiased **P**rediction) has been set up and offers considerable potential in providing information which can predict performance, taking into account ancestors, parents and progeny in different environmental conditions and including a large number of traits. While this requires considerable computer power it does offer scope for improved national coordination as well as more reliable decisions for the individual farmer.

With increasing concentration on a few successful breeds, or with AI, of a few outstanding bulls, there is the potential danger of loss of genetic diversity within the gene pool. Despite these trends, it is encouraging to see interest in helping rare breeds to survive. Two such rare breeds of cattle are the Longhorn and Red Poll. As well as giving pleasure to enthusiasts, these stocks may provide a valuable source of genetic material which could be integrated into the commercial breeding herds. Most breed societies maintain a stock of desirable bulls and a bank of genetic variability in frozen semen. Stocks of rare breeds are important in contributing to this gene bank.

Development of breeds of sheep for wool and meat

In the UK, interest in the selection of breeds of sheep became active in the 18th century. Up to that time, there had been a flourishing wool trade, but only gradual change in the quality of fleece and selection for wool. Then systematic selection for desired traits and detailed recording of matings and performance of the offspring proliferated, leading to the development of today's range of breeds for different purposes. The emphasis has shifted from wool production to meat production, though wool is still produced, almost as a byproduct. In the UK, the modern sheep industry has become highly classified, with breeds suited for certain environmental conditions and chosen according to the intended end product. Increasingly, very detailed computerised records are being kept of individual sheep and their performance in different conditions (see BLUP, described above). This can give valuable information in selecting breeds and individuals for mating.

Selection may be for hardiness (in hill sheep), body weight, lambs born per ewe and survival rate of these lambs; quality of fleece in terms of length, weight or fineness of the wool; rate of growth and quality of the meat and carcass. The actual traits favoured depend on the breed and the location in which the sheep are being reared. At sheep sales and in breed society descriptions, emphasis tends to be on visual characters, sometimes overriding the more elusive but essentially important traits relating to productivity.

A simplified outline of the UK sheep industry illustrates how sheep farmers are able to select breeds to exploit different areas of land which are subjected to a range of weather conditions and, depending on available food, adopt appropriate feeding strategies. From the extremes of pure bred hill sheep to pure bred lowland sheep there is basically a three-tier structure, described as a **stratified system** (Table 4.3 and Figure 4.3). Within this there is systematic cross breeding and movement of sheep from the hill regions through to the lowlands. Superiority and improvement in performance is achieved largely through the cross breeds, with a high proportion of the sales of sheepmeat

Why do you think bulls have a greater influence than cows on the genetic improvement of a herd, on the local and national scale? Think about the number of calves a cow can produce in a lifetime and compare this with the number of inseminations possible from a single bull, even after its death.

coming from cross bred lambs. Movements and breeding of sheep in relation to rearing and feeding are outlined on pages 70–71.

Table 4.3 *Comparing breeds of sheep - some characteristics of hill sheep and lowland sheep and some cross breeds. (Figures in brackets give typical liveweights, calculated as the average of mature ram and ewe weights)*

Hill sheep	Lowland	Cross breeds

Examples of breeds
Herdwick
North Country Cheviot (82)
Scottish Blackface (70)
Swaledale (64)
Welsh Mountain (50)
	Examples of breeds
	• terminal sire breeds
	Dorset Down (77)
	Oxford Down (100)
	Southdown (61)
	Suffolk (91)
	Texel (imported breed) (87)
		Examples of crossbreeds
		Border Leicester × Welsh Mountain
		    = Welsh Halfbred (72)
		Wensleydale × Swaledale
		    = Masham (80)
		Bluefaced Leicester × Swaledale
		    = North Country Mule

• harsh conditions, breed hardy • relatively small body size • lambing delayed until late April or May • normally one lamb per ewe • good milking • good mothering–ability to rear 100% crop • difficult to fatten lambs because of late lambing • good wool weight and quality	• more favourable physical conditions • relatively large body size • lambing from January through to March • high fertility and prolificacy (twins or more per ewe) • good milking ability • good growth rate • fattened to slaughter in 12 to 16 weeks • carcass quality	Cross breeds are bred for a variety of characters–they tend to be intermediate between the hill and lowland though sometimes performance is enhanced

• *longwools – include Bluefaced Leicester (96), Border Leicester (94), Teeswater (96) and Wensleydale (101). Their features combine high fertility and prolificacy with good milking, growth rate and heavy growth of long wool.*

Table 4.4 *Comparing output from hill, upland and lowland sheep in the UK (data in 1991)*

Location	Breeding ewes (thousands)	Finished lambs sold / ewe	Average carcass weight / kg	Total carcass weight / 000t
hill	6692	0.45	15	45 (12%)
upland	3179	1.1	18	63 (17%)
lowland	9251	1.5	19	246 (71%)

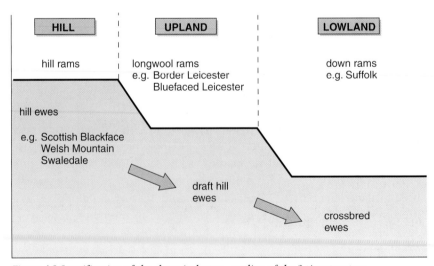

Figure 4.3 Stratification of the sheep industry - outline of the 3-tier structure

Table 4.5 *Comparing average performance of Scottish Blackface and Welsh Mountain ewes under hill and upland conditions (grassland farm)*

Performance feature	Hill environment		Grassland farm	
	Scottish Blackface	Welsh mountain	Scottish Blackface	Welsh mountain
body weight / kg	41	34	67	47
fleece weight / kg	1.7	1.1	2.6	1.8
lambs reared / 100 ewes	81	105	151	136

Reproduction and breeding

Management of the reproductive cycle in cattle and sheep plays an essential part in achieving success in the modern farming industry. Appropriate strategies allow the farmer to

- maintain milk supply throughout the year in a dairy herd
- adjust calving or lambing times to make best use of natural herbage for feeding
- extend lambing season in sheep to take advantage of market demands and favourable prices

Female reproductive system

The main parts of the reproductive system of the cow are shown in Figure 4.4. In the ovary, the **ova** develop from the **oogonia** which arise by mitotic divisions from germinal cells during embryonic and fetal stages. The diploid oogonia first give the primary oocytes, then undergo a reduction division resulting in the production of secondary oocytes and finally the mature ovum (or egg).

The oestrous cycle

The sequence of events in the reproductive cycle are closely controlled by the action of hormones (Figure 4.5). There are two phases within the oestrous (ovarian) cycle, a short **follicular** phase and longer **luteal** phase (Fig 4.7). As the follicle matures, it secretes **oestrogen**, one function of which is to make the cow or ewe come into **oestrus**. This is the time at which she is receptive to the male, shortly before ovulation. A second effect of oestrogen is to stimulate the release from the anterior pituitary of **luteinising hormone** (LH) which is required to trigger ovulation.

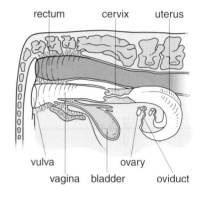

Figure 4.4 Reproductive system of the cow (side view)

- *two **ovaries**, and an **oviduct** leads from each ovary to the **uterus***
- *notice the two horns of the uterus which combine into a relatively short uterine body*
- ***cotyledons** inside the uterus serve for attachment of the developing embryo or fetus*
- *the muscular **cervical canal** separates the uterus from the **vagina** which opens to the exterior at the **vulva***

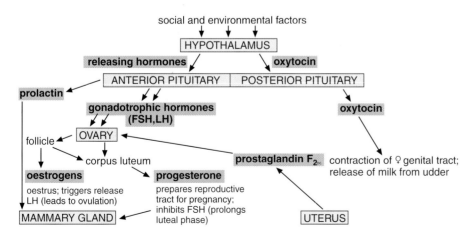

Figure 4.5 - Major hormones involved in the control of reproduction in cows and ewes. In sheep the hormone melatonin is secreted by the pineal gland in the dark. Melatonin has an important effect in the control of seasonality by initiating the hormone cascade from the hypothalamus and anterior pituitary which in turn triggers ovarian activity.

CATTLE AND SHEEP

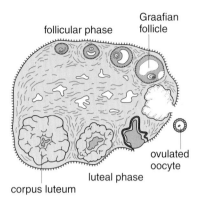

Figure 4.6 Section through ovary of cow or ewe, showing developments during follicular and luteal stages

follicular phase

Graafian follicle

Graafian follicle

ovulated oocyte

luteal phase

corpus luteum

Figure 4.6 Section through ovary of cow or ewe, showing developments during follicular and luteal stages
- *an oocyte becomes surrounded by a single layer of ovarian cells to form a **follicle***
- *cells of this primary follicle then divide to form a multi-layered structure which fills with fluid to become the **Graafian follicle***
- *the Graafian follicle contains the ovum ready for release at ovulation*

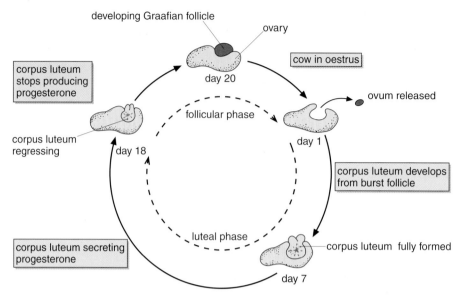

corpus luteum stops producing progesterone

developing Graafian follicle

ovary

cow in oestrus

day 20

ovum released

follicular phase

corpus luteum regressing

day 18

day 1

corpus luteum develops from burst follicle

luteal phase

corpus luteum secreting progesterone

corpus luteum fully formed

day 7

Figure 4.7 Changes in the ovary during the oestrous cycle. The days refer to the cycle in cows – use the text to work out the equivalent days for ewes.
- *the follicular phase is stimulated by secretion of **follicle stimulating hormone** (FSH) and **prolactin** from the **anterior pituitary gland***
- *by day 7 the corpus luteum is fully formed as a solid, almost orange mass of tissue*
- *progesterone prepares the reproductive tract for successful mating and subsequent nourishment of the embryo*

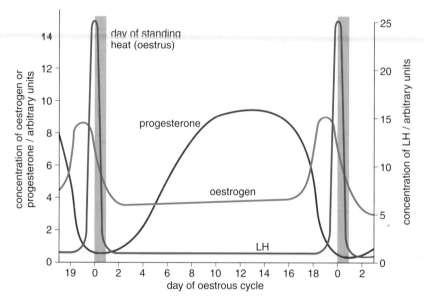

Figure 4.8 Changes in hormone concentrations during the oestrous cycle. The days refer to the cycle in cows. Use the text to work out the equivalent days for ewes.

In cattle, the oestrous cycle typically lasts 21 days. If a cow is not mated with a bull or artificially inseminated, the cycle is repeated throughout the year. Ovulation occurs on day 1 of the cycle. The ovum is released from the follicle which collapses, then undergoes development into a **corpus luteum**. This is the start of the **luteal** phase. Secretion of **progesterone** from the corpus luteum continues for at least two-thirds of the oestrous cycle and is highest between days 11 to 16. Progesterone inhibits the production of follicle

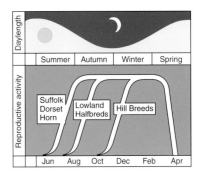

Figure 4.9 Seasonality in sheep - reproductive activity in some breeds in relation to daylength

stimulating hormone (FSH) which would initiate the next follicular stage. This effectively prolongs the luteal phase, increasing the chance of the cow becoming pregnant. If fertilisation does not occur, the corpus luteum regresses (dies back) quite rapidly under the influence of the hormone **prostaglandin F$_{2\alpha}$**, which is secreted by the uterus. Regression of the corpus luteum results in a rapid fall in the level of progesterone, on about day 19, which removes its effect of inhibiting FSH secretion. This then allows release of FSH which stimulates maturation of a Graafian follicle and the whole cycle starts again.

In sheep, the oestrous cycle typically lasts 17 days and is similar to that in cows. The corpus luteum produces progesterone after about two days and prostaglandin, causing regression of the corpus luteum, is produced on about day 11 or 12, allowing the ewe to come into oestrus again after day 17. In non-pregnant ewes, the cycles continue during the breeding season only. The breeding season shows **seasonality** and is influenced by external factors, of which daylength is the most important. In temperate latitudes, the breeding season occurs from late summer and continues to mid- or late winter. This **photoperiodic response** is triggered by shorter daylengths, and is controlled by release of the hormone melatonin from the pineal gland. Secretion of **melatonin** occurs only when dark. Under natural conditions, this seasonal breeding pattern allows the lambs to be born in the spring and at a time when young grass and other herbage becomes available for grazing, giving higher milk production.

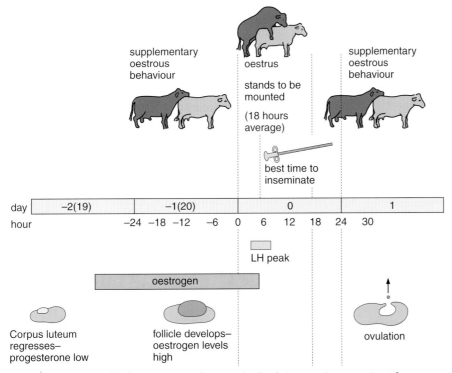

Figure 4.10 Events and behaviour around oestrus in the dairy cow. In cows, signs that indicate the onset of oestrus include
- *a swollen vulva with discharge of watery mucus*
- *she may sniff the vulva or urine of other cows*
- *behaviour includes restlessness, bellowing, licking and mounting of other cows*
- *the cow stands to be mounted – this is the decisive and positive sign of oestrus*

In ewes visible signs of oestrus are less evident.

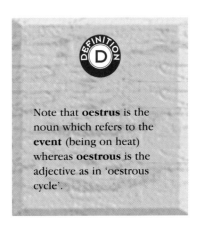

Note that **oestrus** is the noun which refers to the **event** (being on heat) whereas **oestrous** is the adjective as in 'oestrous cycle'.

Oestrus and its detection

The terms **oestrus**, 'on heat' and 'bulling' (for cattle) refer to the time when the cow or ewe is sexually receptive and will stand to be mated (Figure 4.12). Oestrous behaviour generally starts a day or two before oestrus and is linked to hormone activity. In cows the heat period may last between 6 and 30 hours with an average of about 18 hours; in ewes it may last for 24 to 48 hours.

Detection of oestrus is important so that mating or insemination occurs at the best time. Delay results in poorer conception rates or a missed chance for pregnancy, with loss of income while the cow or ewe is being fed, waiting for the next opportunity to be inseminated. In the cow, the best time for insemination to occur is usually between five and twenty hours after the onset of oestrus. Ovulation occurs about 10 to 12 hours after the *end* of oestrus and this timing means the ovum has reached a suitable part of the oviduct. Accurate detection of oestrus is crucial when artificial insemination is used and when cow needs to be taken to a bull in a pen for mating. It is less critical when the bull or ram is allowed to run with the herd or flock.

Artificial control of the oestrous cycle and the timing of oestrus

Controlling the timing of ovulation can give more reliable conception. Synchronisation of oestrus means that a group of heifers would be ready for bulling together or allow artificial insemination (AI) at a pre-arranged time. It may be convenient to keep a batch of cows or ewes together during their pregnancy. They would calve and enter the dairy herd at about the same time and it allows adoption of appropriate feeding programmes during pregnancy. Birth of the calves or lambs occurs at about the same time, and the group can be reared together as a batch through to marketing. With sheep, it is possible to advance the lambing season to capture premium market prices for lamb at Easter and to extend the season to produce out of season lamb. Control of the oestrous cycle may also be used to increase birth of lambs to 3 crops in 2 years or to have a longer season for milking ewes. Disadvantages relate to additional costs, increased handling of the animals and variable performance.

In cows, two approaches are used - both are concerned with the relative length of the luteal and follicular stages of the oestrous cycle. The first attempts to maintain a high level of progesterone in the blood. This prolongs the luteal phase but when the progesterone level falls, the short follicular phase follows leading to ovulation. This can be achieved by inserting into each vagina an artificial device (such as a plastic coil) containing a capsule which releases progesterone thus mimicking days 7 to 18 of the cycle. The hormone is absorbed into the blood through the vaginal wall. When the device is removed the drop in level of progesterone triggers the sequence of events which leads to the development of the follicle. Supply of progesterone can be adjusted so that a group of cows in a herd come into oestrus together. While this method has proved successful in synchronising the timing of oestrus, it may result in lower conception rates in the herd.

The second approach aims to cause regression of the corpus luteum. This regresses naturally at the end of the luteal phase if no mating has occurred and leads into the follicular stage then ovulation. Regression of the corpus luteum can be induced by supplying either prostaglandin $F_{2\alpha}$, a natural secretion from the uterus, or synthetic analogues of this natural hormone. If the hormone is injected after day 7, it will advance the cycle so that in about 3 days the cow is in oestrus. However, in the herd, times of oestrus are liable to be random, so that some cows of the herd may not yet have a fully formed corpus luteum and would not respond to this treatment. A high rate of success has been achieved by giving two injections of the prostaglandin, the second 10 to 12 days after the first. It is then probable that about 90% of the cows have a mature corpus luteum and are likely to respond to the second injection and lead into the short follicular stage. Such cows come into oestrus 2 to 3 days later, showing conception rates similar to that of the normal herd.

In sheep, several methods are used to manipulate the oestrous cycle. As with cattle, synchronisation of oestrus can be encouraged by using progesterone to prolong the luteal phase of the cycle. Sponges, impregnated with a synthetic progesterone, are inserted into the vagina of each sheep in a group at the same time. The sponges are generally left for 12 to 14 days so that this higher level of progesterone lasts longer than that from the natural corpus luteum. Removal of the sponges is likely to lead to a surge of gonadotrophin and ovulation 1 or 2 days later. Pregnant mares serum gonadotrophin (PMSG) mimics the ewe's own gonadotrophins (FSH, LH) and can be injected after the removal of the progesterone sponge, particularly if the ewe was not already cycling when the sponge was introduced. PMSG can also be used to adjust the number of eggs produced.

The teaser effect can be utilised in the few weeks before the expected start of sexual activity in the flock.

Rams – including teasers – give off pheromones which stimulate non-sexually active ewes to ovulate within 2 to 3 days of joining with males.

Teaser rams must be removed and replaced with fertile rams no later than 14 days after the teasers were introduced.

Ewes will respond in one of two ways to teasing:

(1) Show behavioural oestrus and mate with rams some 3 weeks after teaser introduction

(2) Display oestrus and mate some 4 weeks after teaser introduction

Days	0	2	3		14		18	21	28
	Teaser joined with ewes	Ovulation ('silent heat' – no mating)		Teasers out	Fertile rams in		Early mating group		Later mating group

Use one teaser ram for every 40 to 50 ewes	For best results use one fertile ram for every 20 to 25 ewes to be mated

Figure 4.11 Mating in sheep and the teaser effect of rams

Attempts to overcome the seasonality of sheep and extend the breeding season, include timing of the introduction of rams to the flock and use of the hormone melatonin. Introduction of 'teaser' rams into the flock can encourage earlier breeding in sheep which are not yet sexually active. These rams have been vasectomised so are no longer fertile, but still carry the pheromone odours which are attractive to ewes and bring them into oestrus. The teaser rams are left with the flock for about 14 days then replaced with fertile rams. It is likely that the onset of ovulation has been both stimulated and synchronised within a few days.

The seasonality of sheep is linked to the secretion of the hormone melatonin from the pineal gland and this occurs in response to shorter daylengths. This effect can be mimicked by artificial supply of melatonin, resulting in the breeding season being advanced by 4 to 6 weeks. The hormone is supplied as an implant placed subcutaneously behind the ear. This technique should only be applied at certain times during the summer, and the timing depends on the breed. Breeds of sheep differ in their length of breeding season, so some extension can be achieved by cross breeding a breed with a long sexually active season with a shorter one.

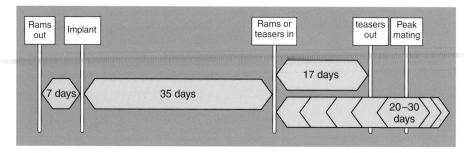

Figure 4.12 Effect of artificial implants of the hormone Regulin on mating activity in sheep.
- *the implant is placed subcutaneously behind the ear*
- *used mid-May to late June for ram introduction in late June and July (Suffolk and Suffolk crosses)*
- *used early June to late July for ram introduction in mid-July to late August (Mule and halfbred flocks)*
- *to be used only at times recommended*
- *rams are separated one week before giving the implant – must be kept out of sight, sound and smell*
- *after treatment rams kept separate for further 5 weeks, then introduced into flock of ewes*
- *very little mating occurs in first two weeks after introducing rams*
- *most mating occurs 20 to 30 days after introducing rams*

Male reproductive system

The main parts of the reproductive system of a bull or ram are shown in Figure 4.13. The sperm pass from the testes along the very long epididymis, which may be over 30 m in length in a bull. The sperm suspension becomes concentrated and the sperms mature and acquire the potential for swimming and fertilising an ovum. Production of sperm can continue throughout the life of the bull from puberty to old age. Rams show seasonality in their sperm production.

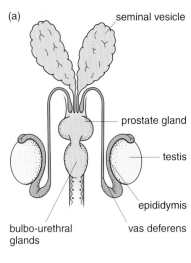

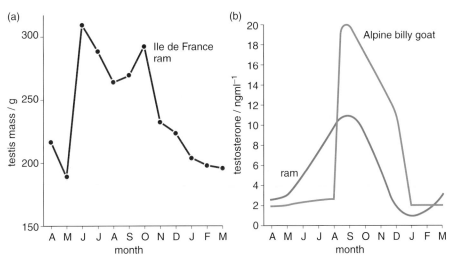

Figure 4.14 Seasonality in rams (data from 49° N): (a) variation in mass of testes (testis mass is adjusted to body mass); (b) testosterone production

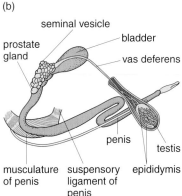

Figure 4.13 Male reproductive system in a bull (a) front view; (b) side view.

Mating and artificial insemination

Puberty, when oestrus is first detected, occurs in cows between 8 and 17 months and in ewes from about 4 months in intensive systems, but much later in extensive situations. The timing varies with breed and is influenced by environmental conditions, especially the level of feeding. Often smaller breeds reach maturity sooner than larger breeds. The presence of a mature bull (or ram) amongst the young females has the effect of bringing on puberty and the farmer may use this to influence the timing of puberty in groups of females. Puberty can be detected by the female showing a degree of sexual excitement and becoming receptive to mounting and mating.

If a bull is allowed to run free with a herd of cows, or a ram with a flock of sheep, mating is likely to occur when the cows or ewes come into oestrus. The bull or ram is stimulated by the odours and secretions from the vulva and may show interest in the urine of the cow or sheep. The bull mounts the cow (or the ram mounts the ewe), the penis is extruded and vigorous thrusting movements by the pelvis follow, which enable the penis to enter the vagina through the lips of the vulva. Ejaculation occurs rapidly and the semen is deposited in the anterior part of the vagina. A bull with a herd can probably inseminate 3 to 4 cows per week, so the ratio of bulls to cows should be adjusted to ensure there is a reasonable chance of successful mating. In sheep, a suitable ratio is about one fertile ram per 20 to 60 ewes to be mated - the higher numbers refer to older rams. To help identify which ewes have mated, the rams are often coloured with a crayon, known as a **raddle**, which rubs off

CATTLE AND SHEEP

For what reasons, other than cost, might a farmer not wish to keep a bull on the farm? What advantages are there in using artificial insemination (AI) instead of natural mating?

onto the ewe when mating occurs. While a system for natural mating may be appropriate for extensive systems, particularly in sheep, there are disadvantages due to irregular conception patterns, uncertainty as to which animals are pregnant and generally less control over the timing of calving or lambing. Many farms now do not keep their own bull, mainly because of the very high cost of bulls with desirable characteristics.

Increasing use is now made of **artificial insemination** (**AI**). Methods used to collect semen are similar in cattle and sheep. With a bull, this can be done either by direct mechanical stimulation of the muscles of the male tract or by exploiting the reaction of the bull to the vagina and its secretions. For the latter, a 'teaser' female cow (one which is maintained in oestrus) may be used to stimulate the bull, then the penis is directed into an artificial vagina which collects the ejaculate. Alternatively a dummy which crudely resembles the shape of a cow is impregnated with urine or smells which encourage the bull to mount it and ejaculate. The artificial vagina is a device which imitates the vagina of a cow thus giving appropriate stimulus to the bull and also maintains a suitable the temperature around the collecting vessel as sperm is highly sensitive to temperature changes.

The collected semen is checked in a laboratory for contamination and for abnormalities. The proportion of active (motile) sperm and their concentration is estimated. A single ejaculate from a bull may produce up to 15×10^6 motile sperm which is far in excess of the number required to give the chance of a successful fertilisation. The semen can be diluted with an **extender** which increases the number of possible inseminations. The extender usually contains

For successful AI, the timing in relation to oestrus is critical. How is oestrus controlled and synchronised in a batch of ewes?

- a sugar which provides energy
- salts which buffer the sperm suspension against changes in pH and the osmotic balance
- a substance such as glycerol which protects the sperm cells against the damaging effects of chilling
- antibiotics which prevent the growth of bacteria

The sperm suspensions are packaged in disposable plastic straws containing about 0.25 to 0.5 cm^3 and stored in liquid nitrogen at –196 °C. There is some slow deterioration of fertilising ability of sperm stored in this way. AI is carried out by inserting a plastic or glass pipette into the cow's vagina (Fig 4.17), guided manually from inside the rectum so that the straw is deposited beyond the cervix.

AI is used less extensively with sheep. Procedures are similar for ewes, though with cervical AI, it is preferable to use fresh or chilled semen rather than frozen. Higher conception rates are obtained using intrauterine AI (also known as laparoscopic AI), whereby the semen is injected directly into the uterus. This is more expensive, both because of the equipment required and the need to have trained people to carry it out.

Important advantages of using artificial insemination (AI) include
- genetic improvement of the stock
- flexibility within the herd or flock in a farm

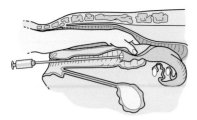

Figure 4.15 Artificial insemination (AI) in a cow. The vagina is manipulated from the rectum to ensure the straw deposits semen beyond the cervix. High rates of conception are achieved provided the original semen is of good quality, that care is taken over temperature changes and the correct time is chosen in relation to ovulation.

- making best use of management strategies in terms of timing of events
- cost effectiveness
- safety
- reduced transmission of disease

Probably the greatest benefit comes from being able to use semen from outstanding sires (bull or ram) in a wide variety of farms, including small ones. This allows the rapid spread of desirable, including new, genetic material locally and on a national and international basis. Deep frozen sperm has been available since the 1940s and its widespread use has made artificial insemination feasible on a large scale. It even allows sperm from a desirable bull to be used after the donor is dead. AI makes it easier, for example, to shift the emphasis in cattle from milk to beef production, and means the farmer does not need to keep several bulls to retain the required flexibility in the breeding herd. By using frozen sperm from a well-tried bull farmers feel more assured of success and consistency in their herd. AI also avoids the time lag waiting for progeny from a young bull to grow, mature and be tested. In this way, one bull can be responsible for several hundred offspring whereas a cow produces only a limited number of offspring in a lifetime. AI can also help in the coordination of a breeding policy on a national scale. In Britain in 1992, about 80% of cattle matings were done by artificial insemination with a high (70%) success rate in relation to conception.

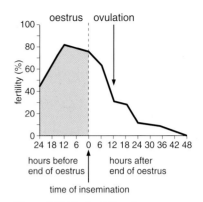

Figure 4.16 The best time for insemination is 12 to 18 hours before ovulation

Fertilisation and pregnancy

Fertilisation normally takes place in the oviduct. The natural motility of the sperm is assisted by contractions of the female genital tract under the influence of hormones, particularly oxytocin. Adrenaline acts antagonistically to oxytocin, so stress or fright may reduce conception rates. As the sperm move through the cervix and uterus towards the oviduct, there is a progressive reduction in numbers from over a million at the site of deposition in the anterior vagina to about one hundred in the oviduct. When AI is used, the straws are inserted well beyond the cervix and deposition of a much smaller number is likely to result in successful fertilisation. During their passage through the uterus, the sperm undergo the final stages of maturation and then can penetrate and fertilise the egg.

The fertilised egg moves along the oviduct and reaches the uterus after three or four divisions. In the cow, attachment of the embryo to the endometrium of the uterus is not completed until about 22 days after fertilisation and occurs by means of a specialised region of fetal **cotyledons** which merge into maternal tissue to form the functions of the placenta. Development of the embryo is similar to that in other mammals. The gestation period in cattle is about 40 weeks and in sheep about 20 weeks. Birth of a calf or lamb marks the onset of lactation or production of milk.

A newly born calf or lamb naturally suckles milk from the mother cow or ewe, then, after several weeks, is gradually be weaned onto solid food. In many farms calves and lambs are fed on artificial milk substitutes, or on milk taken from the cows in the herd, rather than allowing suckling to continue. It is important for the calves and lambs to receive the first milk from the mother, the **colostrum**, within 12 hours of birth and continue for the first few days. Colostrum is beneficial because of its high protein content, especially immune globulins, but particularly for its maternal antibodies, which protect the young calf or lamb against infections likely to be encountered. At this stage, the large protein antibodies in the colostrum can be absorbed directly through the wall of the intestine of the calf or lamb. Colostrum also has an important laxative effect.

Table 4.6 *Composition of colostrum (first 24 h after calving) compared with milk*

Component	Colostrum	Whole milk
Fat (%)	3.6	3.50
Non-fatty solids (%)	18.5	8.60
Protein (%)	4.3	3.25
Casein (%)	5.2	2.60
Albumin (%)	1.5	0.47
Immune globulin (%)	6.0	0.09
Ash (%)	0.97	0.75
Calcium (%)	0.26	0.13
Magnesium (%)	0.04	0.01
Phosphorus (%)	0.24	0.11
Iron (%)	0.20	0.04
Carotenoids / μg per g fat	25 to 45	7.0
Vitamin A / μg per g fat	42 to 48	8.0
Vitamin D / μg per g fat	23 to 45	15.0
Vitamin E / μg per g fat	100 to 150	20.0
Vitamin B / μg per g fat	10 to 50	5.0

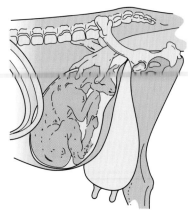

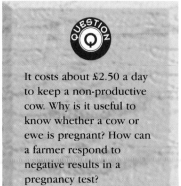

Figure 4.17 Detection of pregnancy by rectal palpation (approaching full term)

It costs about £2.50 a day to keep a non-productive cow. Why is it useful to know whether a cow or ewe is pregnant? How can a farmer respond to negative results in a pregnancy test?

Pregnancy testing

A number of methods are available and are used with varying success (and cost) at different times after possible conception to determine whether insemination has been successful and that the cow or ewe has become pregnant. In ewes, the results of a scan can be particularly useful to detect twins or triplets, so that the feeding programme is adjusted accordingly.

- **rectal palpation** - the arm is inserted into the rectum and the hand can locate and feel the increasing size of the fetus, swelling in one of the horns of the uterus. Later in pregnancy, the fetal head can easily be detected. Used in cows from about 6 weeks (42 days) gestation
- **blood flow in uterine artery** - by the middle of the pregnancy, on the pregnant side of the cow, the pulsations can be felt giving a vibrating sensation, known as *fremitus*.
- **ultrasound scan** - builds up images of the developing fetus from an early stage (30 days in cows). Scans are used in both cows and ewes and are useful because they are highly accurate and do not necessitate the disturbance caused through manual invasion into the cow. This is the main method used for ewes.
- **progesterone** - in the pregnant cow, the level of progesterone is high between days 21 and 24 of the oestrous cycle. The milk is tested for progesterone. The accuracy is 100% for negative results but down to 85% for positive results.

Embryo transplantation

Recent advances in reproductive technology have seen development of techniques for embryo transplantation which could have revolutionary effects when widely available commercially. Ova from a cow or ewe with desirable characteristics are fertilised and allowed to develop to the embryo stage. The embryos are then collected from the donor and transferred to a recipient cow or ewe, in which they develop. The surrogate mother may be a non-pedigree animal, with less valuable characteristics, but provides a means of rearing the embryos before birth.

To maximise the advantage, the donor cow or ewe is stimulated to **superovulate** thereby increasing the number of ova available from the desirable donor. This is generally done by treatment with PMSG, which behaves like FSH and LH, and increases the number of follicles developing at the start of the follicular phase of the oestrous cycle. Several ova are released and artificial insemination with an increased number of sperm then allows these ova to be fertilised and the embryos start to develop. These embryos, ideally aged 6 to 8 days in cows, are then flushed out from the reproductive tract of the donor. This flushing out is possible because of the delay in attachment of the embryos to the wall of the uterus. During this time, the embryo has been floating freely in a fluid environment and can be transferred to a culture medium without harm. These embryos are then inserted into the uterus of the recipient cow, either through the vagina and cervix, or surgically, using an anaesthetic. Embryos are either transplanted into the donor within a fairly short time of being flushed out or they can be stored at –196 °C in liquid nitrogen.

Alternatively, fertilisation and early stages of embryo growth can be carried out *in vitro*. In this method, ova are collected from the ovaries of cows at an

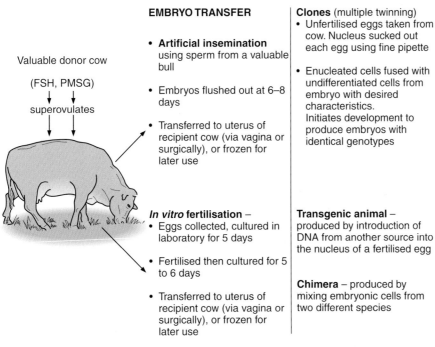

Figure 4.18 Embryo transfer and other ways of manipulating reproduction in cattle and sheep

Find out about the techniques used to produce 'Dolly' the sheep. List some of the advantages of cloning from an adult animal. Cloning may appear to narrow the gene pool, but how could it be used to maintain genetic diversity? What are the potential dangers of widespread use of cloning?

List the hormones which influence reproduction in cattle and sheep. Where is each hormone produced and what triggers its production? For each hormone, list the target(s) it acts upon and then give its effects. In what ways are hormones used artificially to control reproductive events? (You might find it useful to compare your answers with those you would give for human reproduction - see questions, page 152, in *Systems and their maintenance*.)

abattoir. These ova are cultured in the laboratory for about 5 days then fertilised with semen from a bull with desired characteristics. The resulting embryos are further cultured to the five- or six- day stage, then frozen until required for transplantation into the recipient cow. In the cow, the valuable donor can be stimulated to superovulate again after 2 to 3 months, though it is wise to let the cow have a normal pregnancy after two superovulations. This technique has the potential to give a large number of offspring over a few years, compared with an average of one calf a year with conventional breeding.

Embryo transplantation enables a greater contribution to genetic progress to come from the mother. It is likely that embryo transplantation will become available on a wide scale, in both cattle and sheep and has the potential to have a considerable effect on selection and control of characteristics of the herd. A further benefit is that the embryos can be distributed internationally to contribute to breeding programmes, and this is easier than transporting the live adult cows or sheep. Recipient cows or sheep on local farms all over the world could then nurture embryos derived from matings (or artificial insemination) between bulls and cows with chosen desirable characteristics. It is now possible to sex embryos in the laboratory. When this becomes available on a commercial scale, dairy farmers could, for example, choose to implant and rear only female embryos.

Productivity

Feeds and feeding

Feeding is a major input in raising livestock and choice of appropriate feed materials is an important consideration in the productivity of the farm. Nutritive value is determined in terms of its **energy** and **protein** as well as **minerals** and **vitamins**. In formulating feeds, it is necessary to consider its nutritive value in relation to the stage of growth or of the reproductive cycle and to balance this with availability and cost. The diet is adjusted depending on whether the animal is being reared for meat, for milking or production of breeding stock. It is also varied with the age of the animal, its time of birth and the season of year. Feeding levels should be monitored – for example, a high level of feed may result in excess fat being laid down at an early stage in beef cattle whereas a low level could lead to poor wool growth in sheep or poor conception rates in cattle. Improved milk yields can be achieved with better feeding in the early stages of lactation.

Grazing crops of grass, or grass mixed with other herbage, is usually the most economical way of feeding cattle and sheep. However, growth of grass is often seasonal – in temperate climates because the winters are too cold and in other places grass may not grow, for example, in the dry season. The quality of grass varies through the season - young grass has less stalk and a higher nutritive value than that towards flowering. Traditionally, making **hay** has been a means of conserving grass by drying, but increasingly making **silage** (ensiling) is used as an alternative way of conserving grass. Silage may be made with a range of young leafy herbage and under conditions that would be unsuitable for haymaking and with less risk for the farmer.

Other feed materials are listed in Table 4.7. Crops such as kale, turnips or beet can be grazed and are a useful means of providing fresh food at times when grass is poor or unavailable. In this table, the earlier ones are bulky feeds, becoming progressively more concentrated down the list. The concentrated foods are described as being energy-dense or protein-dense – in other words they give a higher level of energy or protein per unit of intake. Concentrates generally cost more than the bulkier foods and this must also be considered when determining rations for the livestock. Compound feeds can be made up from a variety of ingredients, mixed in controlled proportions, with scope for providing high protein levels when required and adding vitamins or minerals. Both hormones and antibiotics have been added to feed for different reasons. Hormones are now rarely included and antibiotics only in certain circumstances, usually with veterinary prescription.

The value of the feed materials is measured in terms of its

- **dry matter (DM)** – the percentage of material remaining after the water has been removed
- **metabolisable energy (ME)** – the energy used by the animal after digestion in its metabolism, deducting losses through gases (methane) and urine
- **digestible crude protein (DCP)** – representing the protein that can be utilised by the animal

Palatability is also significant in determining feed intake. Palatability also affects which species are preferred in a mixed sward of grasses and other herbage and is a reflection of coarseness and sweetness (sugar content) as well as other properties of the different varieties.

Feed materials can be analysed into the energy fraction, protein fraction and ash, allowing suitable rations to be formulated. The diet should ensure that the intake of net energy, protein and other nutrients exceeds the basic maintenance requirement and gives the extra required for production (in terms of meat or milk). Quantities provided are adjusted to match the weight of the animal, the desired weight gain per day or expected yield of milk. In practice, it is difficult to arrive at precise values as any foodstuff will itself be subject to variation. For example, grass will have a much higher energy value in spring (being close to that provided by concentrates) compared with later in the year, and the nutritive value and palatability of silage depends on the time of cutting and the conditions under which it has been fermented and stored. In certain hill regions of Britain the quality of grass is adequate to support cattle reared for beef but not for dairy herds. The cost of grazed grass is about half that of conserved grass and one quarter that of concentrates. During their lives, it is likely that cattle and sheep will be fed a mixture of feeds depending on the rearing system used.

Maize grows better than grass in summer and is now an important crop for silage. Compare the carbohydrate and protein content of grass, hay, and silage from grass and maize (see Table 4.7). What are the advantages to the farmer of making silage for feeding livestock?

In Table 4.7, which of the feeds listed are grown as crops specifically for feeding and which are by-products from other industries?

Table 4.7 *A range of feeds and comparison of their nutrient value – typical (average) values of important features in relation to feeding of livestock.*

Feed material	DM (%)	ME / mJ kg^{-1} DM	DCP / g kg^{-1} DM
grass (grazed)	20.6	11.2	93.0
conserved grass			
hay	85.8	9.2	73.0
grass silage	28.9	10.7	105.0
maize silage	30.4	11.2	70.0
roots			
swedes	10.5	13.9	64.0
turnips	10.0	12.7	70.0
sugar beet pulp (dried)	90.0	12.7	66.0
cereal straw			
barley (spring)	86.0	7.3	9.0
barley (winter)	86.0	5.8	8.0
wheat (winter)	86.0	5.7	1.0
green forage crops			
kale	14.0	11.0	123.0
rape	14.0	9.5	144.0
cereal grains and by-products			
barley	86.0	12.8	82.0
wheat	86.0	13.5	105.0
oats	86.0	12.0	84.0
maize	86.0	13.8	69.0
brewers grains (dried)	90.0	10.4	106.0
other high energy feeds			
molasses	75.0	12.5	31.0
oilseed by-products			
ex soyabean	90.0	13.4	453.0
ex rape seed	90.0	12.0	343.0
legume seeds			
field beans (spring)	86.0	12.8	248.0
peas	86.9	13.0	218.0
animal products			
fish meal (white)	90.0	14.2	631.0
dried skim milk	95.0	14.1	411.0

Explanation of symbols used in the table
DM – or dry matter, represents the material left after all water is removed and is used as the basis for calculating feeds for livestock
ME – the metabolisable energy, or the amount the animal can make use of
DCP – digestible crude protein, or the protein that can be utilised by the animal

Conversion of grass into silage

Silage is fermented grass, effectively pickled in its own juice. The cut grass contains sugars which are fermented by **lactic acid bacteria** under anaerobic conditions to give **lactic acid**. The pH falls to about 4.0 which is low enough to prevent spoilage of the product by other microbes. When ready to make silage, the grass and other herbage is cut and wilted for several hours. This allows it to lose some of its mass, mainly as water, before being transferred and packed closely into the silo. The container can be in the form of a pit, in clamps, towers or now often it is done in big polythene bales. The fermentation to lactic acid is due mainly to species of *Lactobacillus* and *Enterococcus*, which are already present amongst the microflora on the harvested grass. Respiration of the plant material continues even after cutting, using up available oxygen. Provided the silo is well sealed, conditions soon become anaerobic which encourages the activity of the lactic acid bacteria. Temperatures may rise inside the silo, but provided the rise is not excessive, the warmth encourages the fermentation processes. As lactic acid is produced, the pH falls, to between 4.2 and 3.8. By this time, the grass has been converted to silage, which is dark green in colour, has a sharp acid taste, little smell and if kept in suitable conditions is stable for several years.

As well as the bacteria which produce lactic acid, other undesirable bacteria are present which may affect the progress of ensiling. Under cold and wet conditions, with low sugar availability, anaerobic species of *Clostridium* convert sugars and lactic acid to butyric acid. This results in a rancid smell and makes the silage unpalatable. Species of *Enterobacillus* convert sugars to acetic acid. Butyric and acetic acids are less effective in lowering the pH, so it takes longer

Suggest advantages to the farmer of making silage in bales.

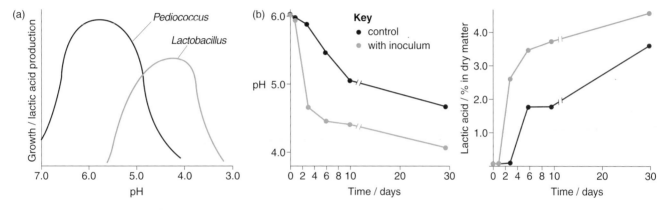

Figure 4.19 (a) Making silage. Different bacteria contribute to production of lactic acid and lowering of the pH in fermentation to produce silage. The action of natural microbes can be enhanced by adding microbial inoculants of suitable bacteria; (b) two graphs showing the effect of inoculants on the rate of lactic acid production and pH change in silage made from vetch.

Figure 4.20 Making silage. The cut grass is packed closely into the pit. The polythene cover, weighted down with old car tyres, helps to maintain anaerobic conditions inside.

for the pH to fall to 4.0. More energy is lost in the process giving less carbohydrate for feeding the animals. Both *Clostridium* and *Enterobacillus* convert proteins to ammonia, which raises the pH, so even more sugars must be broken down to achieve the desired pH. If the silage is exposed to air, aerobic species may oxidise carbohydrate material to carbon dioxide and water.

- Why is cutting of grass for silage usually done after midday and why is it best to cut for silage on sunny rather than on cloudy days?
- Cutting and bruising of the grass helps to release juices and increases silage density. How does consolidation and dense packing of the silo affect the silage processing?
- A commercially produced inoculant supplied to farmers had the following description: *Lactobacillus plantarum*, *Pediococcus acidilactici*, *Streptococcus faecium*, *Lactococcus lactis lactis*, Clostridial bacteriophages, hemicellulase and cellulase enzymes. Suggest the likely benefit of each of these in making good quality silage.

CATTLE AND SHEEP

QUESTION

At different stages in the silage making process, there is some reduction in sugar content, which means a loss of food reserves for feeding the stock. Identify where these losses occur and whether this can be prevented.

The quality of silage depends on the initial mixture of grass or other herbage and on the conditions under which it is grown and harvested. To get the best silage during processing, the low pH should be reached quickly with minimum loss of carbohydrate reserve. Various procedures are adopted to help achieve this, including the use of **additives**. Generally these aim to encourage the desirable anaerobic bacteria at the expense of the undesirable. The sugar content of the crop should be high at the time of harvest. This can be improved by selection of suitable grasses and by harvesting after midday, in sunny weather. **Wilting** helps reduce the quantity of material to be pickled so less lactic acid need be produced. **Cutting and bruising** the crop helps to release the cell contents and encourages rapid lactic acid production. The cut grass should be as clean as possible to avoid contamination by the *Clostridia* and other species which are present in soil. Sugar in the form of **molasses** is often supplied to increase the water soluble carbohydrate available for the bacterial population (*Lactobacillus*). The liquid molasses can be added by spraying to ensure good mixing with the silage material. **Enzymes** such as cellulases and hemicellulases are used to break down cellulosic material and so increase the available soluble carbohydrate. Addition of **preservatives** (such as formalin) with or without **acid** (such as formic acid) inhibits activity of *Clostridia* species and provides a suitable pH for the *Lactobacillus*. **Inoculants** of desirable species of bacteria, particularly *Lactobacillus* and *Streptococcus* are often applied to ensure rapid fermentation to lactic acid.

Grassland and grazing

In the UK, grasslands fall into three main types:

- **rough mountain and hill grassland** – the plants have rather low food value and short growing season. When used for sheep and beef grazing, the stocking density is relatively low.
- **permanent pasture** – remains unploughed for long periods so a good turf becomes established and withstands treading in wet weather. This can support a wide range of grasses, usually native species, and its quality depends on the species composition.
- **leys** – are ploughed and resown annually, or may be left up to about 4 years. Selection of appropriate grasses, clovers or other species and application of fertilisers can give a productive pasture, but it is more liable to damage through treading

Grazing animals are selective in their choice of species. New shoots, known as **tillers**, continue to be produced from the base of grasses throughout the growing season. Tillering is stimulated by grazing and this provides fresh new growth, until the plant switches to producing flowerheads. At that stage, the food value of the grass is very much less.

Sheep nibble vegetation close to the ground and thereby maintain a short, even sward. Cattle, however, wrap their large tongue around the vegetation and, compared with sheep, can consume relatively tall and coarse plants. The resulting sward is more uneven than with sheep because cattle tend to select certain patches. They also trample, leaving hoof marks, large cowpats, bare soil and muddy areas. Sheep use their bottom teeth against the horny dental pad

QUESTION

In Britain, the main domesticated animals are sheep, cattle and horses. What influence have these, together with wild animals (particularly rabbits), had on natural ecosystems? Think about effect grazing has on the nature of grassland and in prevention of the succession of grassland to scrub. What would be the climax community if there were no grazing? How far can grazing become integrated into conservation strategies?

to cut the grass off the plant. In terms of selection, their preference is firstly for herbs and grasses, then for sedges and dwarf shrubs, in that order. The ideal sward height for sheep is between 2 and 6 cm. Sheep tend to avoid the least desired species, so at certain times it may be advantageous to increase the stocking density to ensure these are cleaned from the pasture and not allowed to flower and so dominate the other species. Cattle tend to pull the herbage by wrapping their rasping tongue around the leaves or stalks and cutting it between their lower teeth and upper dental pad as they swing their head. The ideal sward height for cattle is between 5 and 12 cm - if shorter, they require more bites, hence a longer grazing period to take in the same amount of material. Cattle can ingest flower heads to a greater extent than sheep, so help to maintain a more diverse species composition in the sward.

Digestion and the alimentary canal

Cattle and sheep are herbivores. Their ruminant digestive system allows a high proportion of the cellulose in fibrous foods to be digested and become available as energy. They also benefit from features of their nitrogen metabolism and their ability to synthesise water-soluble vitamins. The alimentary canal is more complex than that of humans due to three additional compartments, the **rumen**, **reticulum** and **omasum**, which come before the true stomach, known as the **abomasum** (Figure 4.22).

After swallowing, food passes to the rumen for up to 30 hours. Here it is mixed mechanically and fermented by populations of microorganisms. Coarse material is regurgitated into the mouth, rechewed, then swallowed again. Chewing the cud in this way may continue for up to 8 hours in a day if the diet is very fibrous. The microorganisms in the rumen are mainly bacteria but also include protoctists and yeasts. These microbes become established soon after birth when the calf begins to pick up solid food. The species mixture depends on the food consumed so changes to the diet should be made gradually to ensure rumen microorganisms adjust accordingly. The microorganisms digest carbohydrates, particularly polysaccharides with b-links, and this contributes to the breakdown of cellulose. The resulting hexoses are further broken down, by fermentation under anaerobic conditions, to short-chain organic acids (ethanoic, propionic and butyric) with the release of the gases carbon dioxide and methane.

example of rumen fermentation reaction
$C_6H_{12}O_6 \rightarrow 2CH_3COOH + CO_2 + CH_4$

Energy released in these reactions is utilised by the microorganisms for their own biosyntheses. The gases escape from the animal by belching and are wasted. The acids, known as volatile fatty acids (VFAs) are absorbed through the walls of the rumen and contribute to the energy requirements of the animal. High fibre foods rich in cellulose produce mainly ethanoic acid, whereas feeding of concentrates increases the proportion of propionic acid produced. In dairy cows, a shift towards more propionic acid results in deposition of body fat with a reduction in fat content of milk and a lowered milk production. The composition of the rumen microflora can be altered artificially so that the carbohydrate, when broken down, produces more or less methane.

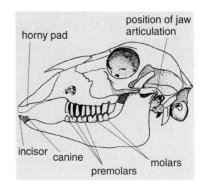

Figure 4.21 Side view of skull showing dentition of a sheep, illustrating typical herbivore features.

- small sharp incisors and canines on lower jaw – cut against pad of gum for cropping
- diastema (gap between incisors and molars) – for manipulation of food bolus
- premolars and molars – for grinding
- ridges of enamel – provide efficient cutting edge as grinding surfaces of premolars and molars get worn down
- open roots – allow for continuing growing of the tooth
- loose jaw articulation – allows sideways movement of lower jaw against upper jaw (related to cud chewing)

Dental formula $\quad$ i $\frac{0}{3}$ c $\frac{0}{1}$ pm $\frac{3}{3}$ m $\frac{3}{3}$

(i = incisors, c = canines, pm = premolars, m = molars. The line separates the upper and lower jaw)

CATTLE AND SHEEP

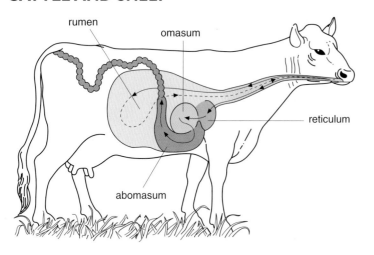

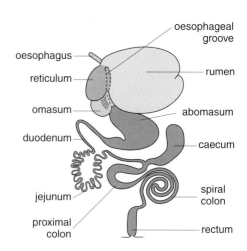

Figure 4.22 Ruminant digestive system in a cow– the arrows show the pathway taken by food, including chewing the cud

- *the **reticulum** has no digestive function – concerned with the passage of boluses of food to the oesophagus and digested material from the rumen to the omasum*
- *the **rumen** is the chamber in which fermentation of food material takes place. Capacity is about 150 litres. Microorganisms in the rumen synthesise vitamins of the B-complex so it is unnecessary to supply these with food*
- *the **omasum** – main function is to remove water and organic acids from digested material passed from the rumen; produces no digestive secretions. Capacity is about 15 litres.*
- *the **abomasum** functions as a 'true' stomach, secreting gastric juices. Digestion here and in the rest of the alimentary canal is similar to that in humans*
- *fibre slows down the passage of food in the gut and adequate amounts in the diet are essential to maintain rumen function.*

Gas production from carbohydrates fermenting in the rumen can account for loss of about 7% of the energy contained in feeds. Why might it be desirable to reduce the methane production? Think about the diet and economics of feeding the cow or sheep as well as about wider environmental implications.

Protein is also broken down in the rumen by microbial activity, first to amino acids then deaminated to release ammonia. Some ammonia is incorporated into microbial protein and the rest is absorbed into the blood of the animal. It may then either be excreted or recycled to the rumen by means of the saliva and thereby gain another chance of being synthesised into microbial protein. Non-protein nitrogen, in the form of chemicals such as urea, can be supplied with the diet and is utilised by microorganisms to synthesise protein. This protein becomes available to the animal when they pass from the rumen into the abomasum and are digested. Some of the protein in the food by-passes the rumen and is then digested with the animal's digestive enzymes in the abomasum, duodenum and rest of the alimentary canal.

Figure 4.23 Ruminants and their utilisation of nitrogen in the diet. Note how ruminants benefit from microbes in the rumen which convert non-protein and low-quality protein into relatively high-quality protein. NPN – non-protein nitrogen, such as urea, RDP = rumen degradable protein, UDP = undegradable protein

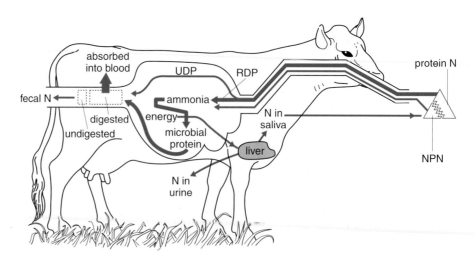

Rearing systems for cattle – for beef or milk

Female calves are kept for milk production and to give birth to more calves, and some are reared for beef. Male calves are reared for beef though excess male calves are used either for veal or are slaughtered at birth. Some calves are reared on the farm where born, but often they are transported to other areas and 'bought in' specifically to rear for beef production. In the UK, calves are born throughout the year. Feeding systems aim to make best use of the summer growing season, alternating with conserved foods in different forms, with or without concentrates.

With **intensive systems**, the animals are housed or kept in yards, sometimes throughout their lives. In **semi-intensive** to **extensive** systems, they spend a greater proportion or all of their lives grazing on grasslands. **Housing** gives protection against adverse weather conditions and allows close control over feeding. The housing should be light and airy – adequate ventilation helps to reduce chance of disease. For cattle, usually no control over temperature is required, though greater protection is required for young calves. A good layer of straw or other bedding is essential, for insulation and to help with the cleanliness, but there must be adequate provision for disposal of litter and draining away of slurry (liquid waste from the animals).

Intensive systems for beef keep cattle housed throughout their lives, on a diet largely of cereals with or without silage. High daily weight gains are achieved (1.25 kg per day) with the animal being slaughtered after about 11 months at a weight of 430 kg or more. Costs of providing food are higher and there is increased risk of disease.

In **semi-intensive** systems for rearing beef in Britain, calves born in the autumn (September to December) are likely to be slaughtered between 16 and 20 months, depending on breed. During the first winter they are fed a mixture of forage and cereals and may gain about 0.7 kg per day reaching a weight of 200 kg by the spring. They are turned out on to grass for the summer, and, with an increased daily weight gain, reach 360 kg by October, when they are returned to yard enclosures. Here they are fed on a high nutritional level, to be 'winter finished' for slaughter and sale before the spring. Calves born in winter are likely to be older at slaughter, but this longer slower growing period results in a heavier, leaner body weight with lower consumption of expensive cereals. They have two summers out on grass - by the end of the first, weights may reach 330 kg. During the winter, a relatively low level feed of silage with minimal supplements of minerals and vitamins is given. During the second summer, the cattle are again turned out to grass, achieve high growth rates of over 1 kg per day to reach a weight up to 600 kg. Table 4.8 summarises feeding systems for beef cattle.

Levels of feeding are more critical in a dairy herd where high daily yields of milk are expected. In the UK, most dairy cattle graze grass in the summer though many receive additional concentrates at the time of milking. In winter the bulk of the food comes from hay, silage or root crops, supplemented by concentrates. All cows should have adequate access to the feeds but avoid an excess of concentrates over fibrous foods which could depress appetite or

CATTLE AND SHEEP

Table 4.8 *Feeding systems for rearing beef cattle in the UK. The figures represent targets, from three-month old calves, though actual weights at slaughter depend on breed. The lower end of the range gives figures for Hereford × Friesian or Holstein × Friesian whereas the higher values would be achieved by Charolais or Simmental × Friesian.*

Rearing system	Main feed materials	Approx. age and weight at slaughter	Daily weight gain
cereal beef	fed cereals, such as barley, with protein, vitamin, mineral supplements	11 to 13 months 460 to 520 kg	~1.3 kg
grass silage	mainly grass silage with some supplements; allows high stocking rates	14 to 17 months 480 to 500 kg	~1.0 kg
maize silage	mainly maize silage with some supplements; favoured in Europe where good maize harvest	12 to 16 months 450 to 525 kg	~1.1 kg
grass, finished with silage / cereal	*autumn born calves* – grazed from 6 to 12 months, finished on silage / cereal	16 to 20 months 475 to 550 kg	~0.8 to 0.9 kg
grass beef	*autumn and winter born calves* – grazed in first summer, fed store rations in second winter, grazed to finished in second summer	20 to 24 months 500 to 610 kg	~0.8 to 0.9 kg

reduce rumen function. There is a close correlation between energy intake and milk yield, which means that higher levels of nutrition are required earlier in lactation when output is high.

Rearing systems for sheep

An outline is given on page 50 and Figure 4.3 of the stratified system in the UK in relation to breeds of sheep. Here we look at the rearing and feeding of sheep in the **hill areas** (over 500 m), **uplands** (300 m to 500 m) and **lowlands** (under 300 m), with an indication of the reasons for movements of sheep between the regions and the complex system of breeding and cross breeding at different stages in the lives of the sheep.

In the annual cycle for **hill sheep** feeding and lambing are closely integrated with the normal seasonal changes. Such systems are **extensive** with relatively large areas of relatively poor grassland available for grazing at a low stocking rate. Hill sheep are well adapted to these harsh conditions and usefully exploit the grazing in these areas. At lambing time, ewes and their lambs may be given improved grazing and protection closer to the farm. Ewes are generally kept on the hills for 2 or 3 breeding seasons. As they start losing condition, they are then sent (**drafted**) to upland farms where they are cross-bred with lambs from other breeds. In the better upland conditions, the breeding life of hill ewes can be extended for a few more seasons.

In **upland** farms, the conditions are less harsh - improved grazing allows a higher stocking rate and higher fertility can be achieved. Upland farmers use their local upland breeds (such as longwools) to cross breed with the hill ewes. The longwools contribute desirable genetic characteristics, such as increased size and growth rate, prolificacy (seen as twin lambs rather than singles) and good milk yield. The ewes of these cross breeds are generally sold to lowland farms for further cross breeding whereas the males are sold either for meat or on to lowland farms where they can be fed on winter root crops to reach maturity.

A list of 'freedoms' in relation to welfare, is given on page 93, Chapter 5, (Chickens). How far do intensive systems used with cattle and with sheep, conflict with these freedoms in animals denied access to pasture? What are the benefits, to the farmer and to the animal, of intensive systems and what are the disadvantages? Intensive systems have become established in the farming industry in many parts of the world, but how far would it be practicable to return to extensive systems on a global scale?

Lowlands are favoured for arable farming and for dairying, but sheep can also be profitable in these areas because they are relatively cheap to maintain. Lowland sheep are reared mainly for meat. They are grazed on pastures improved with fertilisers at a high stocking rate. The farmer aims to produce at least 160 lambs per 100 ewes. The rams used to produce these lambs are known as **terminal sires** and are selected for their genetic characteristics of carcass leanness, high meat to bone ratio as well as good growth rate and prolificacy. Lowland farms can continue to make use of grazing in winter, though the nutritional value is lower. Feeding is usually supplemented by conserved crops, such as hay or silage.

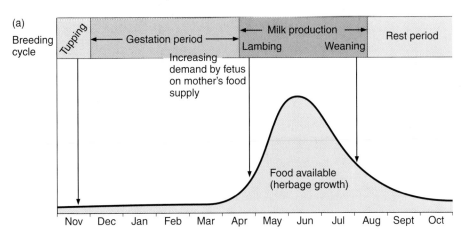

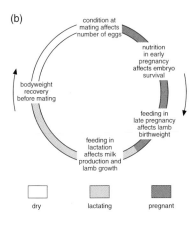

*Figure 4.24 (a) The annual breeding cycle of hill sheep is closely linked with seasonal changes and available feed; (b) the annual cycle of feeding and production in sheep. Ewes are often given an improved level of feeding in the 3 to 4 weeks before mating. This practice, known as **flushing**, has been part of traditional sheep farming for centuries. Flushing improves the body condition of ewes and increases the rate of ovulation, hence the fertility of the flock.*

Control over diet with sheep is probably less precise than in cattle, as the latter are fed predominantly on concentrates and compound feeds when under intensive conditions. Winter feeding of sheep may include forage root crops, and there is some intensive production, giving cereals or compound feeds. The sheep must then be housed and careful consideration needs to be given to the profitability.

Growth and meat production

Growth of an animal during its life is seen as increase in mass as well as changes in form and composition. As cattle and sheep grow, nutrients contribute preferentially first to bone, then to muscle and finally fat. Thus a young animal has a higher proportion of bone in its body mass and a mature animal has greater deposition of fat. The breed can affect both the distribution of muscle and the rate at which growth occurs. Compared with early maturing breeds, late maturing breeds show a higher rate of liveweight gain linked to a larger skeletal size and muscular development, with fatty tissue being laid down at a later age. Adjustment of feed can, to some extent, control the rate of weight gain and

Figure 4.25 (top) A typical hardy mountain sheep; (below) a collection of lowland sheep, housed in a polytunnel around lambing time. The fencing allows lambs, but not adult sheep, to reach extra food

ratio of muscle to fat. Meat is derived from the muscle of the animal, and the aim of the farmer is to produce a carcass with a high proportion of desirable lean meat cuts, with minimum bone in relation to muscle.

Many natural hormones contribute directly or indirectly to metabolism and growth. Males generally grow faster than females. Thus in cattle, a young bull grows faster and is leaner than a heifer (young female) at the same age. The effect of natural hormones can be illustrated by castrated males (known as steers) which are deprived of their supply of androgens. In their metabolism, the energy is diverted into synthesis of fat rather than muscle with the result that a young bull of the same age produces a carcase which is 10% heavier but still leaner than that of the steer.

Hormones can be used artificially to manipulate growth, though this practice has led to controversy. As a result of consumer concern over possible residues in meat, their use was banned in 1986 in the EEC and they are not currently used in the UK. Steroid hormones can be implanted under the skin of the ear and such cattle show an increase in daily gain and leaner carcasses. Generally the best responses in males are with oestrogens and with androgenic substances in females. To a lesser extent, use of the growth hormones bovine somatotrophin (BST) also stimulates leaner and more efficient growth. A future possibility for manipulation of growth and development may lie in an immunological approach in which cattle are stimulated to produce antibodies which interfere with the release and interaction of different growth hormones.

Milk production in cows

Milk production and the calving cycle

On the farm, a suitable interval for calving is about 365 days so we can look at the events in a breeding cow on an annual basis, for both beef and dairy cows. The length of each reproductive cycle can vary, depending on the season, feeding strategies, deliberate manipulation of the breeding, individual characteristics of the cow and the breed.

A **lactation** starts at the birth of a calf, and to some extent its length can be adjusted according to the desired calving frequency. An ideal length would be 305 days, followed by a 56-day dry period. The peak yield of milk production occurs between 4 and 6 weeks after calving, then falls steadily. A good dairy cow in a modern milking herd is likely to yield between 6 000 and 7 000 litres of milk in a year. The **drying off period** allows the cow to recover physiologically from the heavy demands of producing milk. If the cow is to calve again at the end of the dry period, we need to look at the beginning of the cycle to determine the appropriate time for the cow to be mated. The earliest time after calving that the cow would come into oestrus is at about 42 days, with successive oestrous cycles following at 21 day intervals. The actual length of the calving cycle then depends on when the cow becomes pregnant. A 365-day cycle includes an 82-day interval between calving and mating. In practice, in many herds a cow produces a calf about every 390 days.

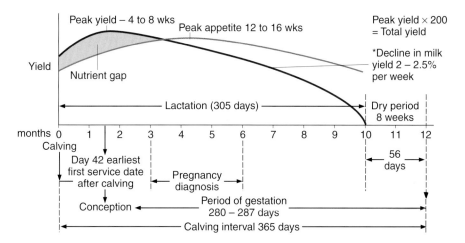

Figure 4.26 Events during the annual breeding cycle in a dairy cow – with a convenient 365-day calving interval.

The yield within a lactation reaches a peak in the first few weeks then falls steadily (Figure 4.31). Yield is also influenced by the time of year at which calving occurs, the age of the cow and particularly by the feeding strategy. In Britain, cows are often turned out onto grass in spring, after wintering indoors, and the effect of young grass and other foliage can result in a 10 to 15% increase in daily yield. The daily yield can also be influenced by changes in the frequency in milking - increases of 10 to 15% can be achieved by milking three times a day (instead of the usual twice daily milking), provided adequate feed is available. Lactation can be prolonged by continuing to milk the cow, but daily yield begins to fall off sharply when the cow becomes pregnant again. There is variation between breeds, the highest yields (over 10 000 litres) being achieved from British Holsteins. Guernsey and Jersey cows yield much less, but their milk is valued for its higher milk fat. In 1991 in Britain, the average yield per cow was 5 200 litres, compared with an average of 3 100 litres in 1956. This improvement can be attributed to selection and genetic improvement in the milking herd and also to management of feeding.

The cow has four **mammary glands**, contained in the **udder** which shows considerable increase in size during the later stages of pregnancy and lactation. Selection for high yields has led to grossly enlarged udders which can be awkward for the cow when moving and cause problems if inadequately supported by the suspensory ligaments which attach the glands to the body wall and pelvic girdle. In breeding dairy cows, there is selection for suitable **conformation** of the udder as well as its shape and positioning of the teats, otherwise there may be difficulties in applying teat cups during milking by machine. The condition known as **mastitis** is an inflammation of the udder, due often to bacterial infection but also resulting from physical damage to the tissues. Mastitis leads to lower milk yields. Some cows show greater susceptibility than others to the condition so this is another character to be considered in selection for a milking herd.

Milk is produced by the milk-synthesising cells in the mammary glands, using simple nutrients in the blood. The protein casein and the sugar lactose are

If a cow is not pregnant by third cycle, a farmer may need to take action. How is oestrus detected and how can it be synchronised in a herd? How is pregnancy testing carried out and why is it important to know if a cow is pregnant? What action can a farmer take to increase the likelihood of successful insemination?

Why does calving frequency vary? Which month in the lactation does a cow become pregnant again?

Which hormones stimulate the development of mammary tissue during pregnancy?

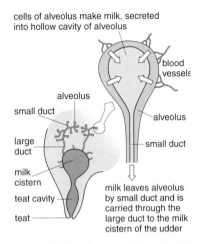

cells of alveolus make milk, secreted into hollow cavity of alveolus

blood vessels

alveolus

small duct

alveolus

large duct

small duct

milk cistern

teat cavity

teat

milk leaves alveolus by small duct and is carried through the large duct to the milk cistern of the udder

Figure 4.27 Synthesis and flow of milk in the udder, with enlarged view of an alveolus.

found naturally only in milk, whereas other substances pass unchanged from the blood into the milk. There is high demand on nutrients - about 500 dm³ of blood pass through the mammary glands in the production of 1 litre (dm³) of milk. The newly formed milk is secreted into cavities (alveoli), then passes into a series of ducts and cisterns leading towards the teat. It is stored until released from the teat, either by the suckling calf or when milked.

Release of milk from the udder is a complex process, under both nervous and hormonal control. Sucking or mechanical stimulus of the teat results in a nervous reflex in the brain which leads to the release of the hormone oxytocin from the posterior pituitary gland. This stimulates a rush of milk along the ducts, and a build-up of pressure behind the teats. Nervous control activates a series of events, culminating in the opening of the sphincter in the teat to allow the milk to flow out. The oxytocin may also be released in response to stimuli received through the eyes, ears or nose. A familiar and undisturbed milking routine is important if successful release of milk is to be achieved in a relatively short time. Machine milking of cows has become highly efficient, taking about 6 minutes per cow.

Factors which influence the composition of whole milk include breed, age, feed, health (of cow), season and stage of lactation (see Figure 4.31). The average composition of whole milk from Friesian cows is compared with that from Channel Island (Guernsey and Jersey) cows in Table 4.9.

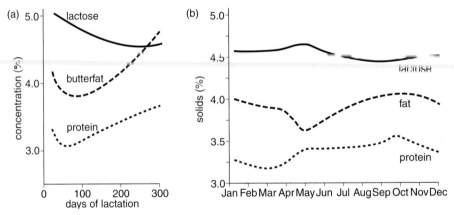

Figure 4.29 (a) Changes in composition of milk during a lactation (lactose, butterfat, protein). Butterfat and protein contents are high for the first few days of lactation, dropping towards about week 12 (during which time the yield is at its highest). They then increase during the drying-off period, and by week 44 return close to the level achieved at the beginning of lactation. Lactose content remains about the same throughout lactation; (b) variation in milk constituents with season.

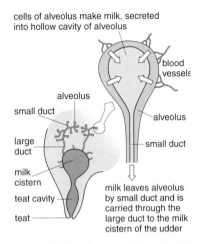

Figure 4.28 A typical milking parlour consists of a series of adjacent stalls, which may allow the cow to be fed concentrates during the milking. The cow enters the stall, the udder is washed and the cluster of teat cups of the machine is attached. The machine, by its pulsating suction, both stimulates the flow of milk from the teat and withdraws it into the collecting jar. The milk from each cow is recorded individually. It is essential for all equipment to be kept clean by sterilising after each milking.

Quality control of milk

Milk from most farms in Britain is collected in bulk in tankers and transported to a processing dairy or manufacturing unit. Milk intended for consumption as liquid is treated to reduce contamination by microorganisms and prolong its shelf life. The quality of milk produced is carefully monitored, with respect to both composition and standards of hygiene. In the UK, samples of milk are taken from all farms and subjected to a range of routine laboratory tests. Results from these tests define the quality of the milk and may be linked to payments (or deductions) made for the milk from a particular farm.

The **composition** can be monitored by measurement of the infra-red absorption by the milk sample to give a direct digital reading of **lactose**, **protein** and **butterfat** content. Bacteria in the milk can be estimated by mixing milk with nutrient agar and incubating this at 30 °C for 3 days. The number of colonies are counted automatically and recorded on a computer and give a measure of the **total bacterial count** (known as **TBC**). Presence of bacteria can be an indication of mastitis inside the udder or arise as contamination outside the udder, from dirty teats or milking equipment. An indication of the level of mastitis is given by the **Cell Count Test**. In this test, most of the cells counted are white blood cells which increase in number in response to disease, including presence of mastitis-causing bacteria in the udder. Another test checks for brucellosis in the cow, by detecting **antibodies** in the milk. Milk should be free from residual **antibiotics** which may have been given to cows suffering from disease, such as mastitis. Presence of antibiotics in milk is undesirable because they may lead to development of resistant bacteria in consumers. Tests to detect antibiotics involve growing sensitive bacteria on nutrient agar with a small quantity of milk. An indicator dye (brom cresol purple) is incorporated in the growth medium. If antibiotics are present in the milk, growth of bacteria is inhibited and the indicator remains blue. Growing bacteria produce acid which turns the indicator yellow.

Other tests check the effectiveness of pasteurisation, sterilising and ultra heat treatment. The **phosphatase test** detects whether the enzyme phosphatase, present in milk, has been inactivated during the pasteurisation process. Phosphatase is slightly more resistant to heat than the bacterium *Mycobacterium tuberculosis*. The chemical disodium *p*-nitrophenyl phosphate is added. If active phosphatase is present, this chemical breaks down, producing is *p*-nitrophenol, which is detected because it gives a yellow colouration. A **methylene blue** test can check the keeping quality of pasteurised milk. Sterilised milk is checked by the **turbidity test**. When milk is sterilised, certain proteins are denatured so their structure changes. Ammonium sulphate is added which leads to precipitation of denatured proteins. the filtrate gives a clear solution which remains clear on heating. Turbidity indicates that sterilisation has not been carried out effectively.

Pasteurisation and sterilisation of milk

In the UK, about 93% of milk is pasteurised, and the rest is either ultra heat treated (UHT) or sterilised. Only a fraction is sold as unpasteurised milk. In pasteurised milk, the flavour and nutritional content is hardly affected by the pasteurisation process. **Pasteurisation** aims to kill pathogenic organisms and to reduce the number of other non-pathogenic bacteria which would cause spoilage. Pasteurisation is not effective against spores. In particular, pasteurisation is expected to make the milk safe from *Mycobacterium tuberculosis*, and from *Brucella abortus*, the causative organisms of tuberculosis and brucellosis respectively. Pasteurisation extends the keeping time of the milk for a few days, provided it is kept refrigerated (between 1 and 4 °C). **Sterilisation** uses heat but at a higher temperature than pasteurisation. Sterilisation aims to destroy bacteria and other microorganisms which may not be killed by pasteurisation, though some very resistant spores may survive the process.

In some countries a system of milk 'quotas' has been introduced in an attempt to regulate the production of milk. Farmers who exceed their allowed quota in terms of volume of milk and or in its composition, may incur financial penalties. How can a farmer control or adjust the milk produced - on the farm as a whole and per individual cow?

How would antibiotics in the milk affect production of cheese and yoghurt?

Pasteurisation is carried out either by heating the milk at a temperature of 62.8 to 65.6 °C for at least 30 minutes, or by heating to 71.7 °C for at least 15 seconds. This 'high-temperature short-time' (HTST) system is now widely used. The milk is passed through pipes or between plates in a heat exchanger system, surrounded by hot water kept at a temperature just above the required temperature. It is carefully controlled to ensure the milk is held at the correct temperature for the appropriate time, then rapidly cooled to about 3 °C.

Temperatures used for sterilisation are at least 100 °C, though often the milk is sealed into bottles and heated to a temperature of approximately 115 °C for 20 minutes, then cooled rapidly. Sterilised milk should keep for several months without refrigeration, but the flavour is noticeably altered giving a caramelised taste. It has also become homogenised with loss of separated cream. UHT milk (ultra-high-temperature sterilisation) is prepared by heating milk to at least 132.2 °C for at least one second. UHT milk is sealed in cartons, aseptically packaged in sterile conditions. UHT milk has a keeping time of several months with flavour qualities close to that of pasteurised milk.

QUESTION

Why can milk go sour even after it has been pasteurised?

Table 4.9 *Nutrient quality of different milks*

| Nutient | Type of milk – composition as g per 100g | | | | |
| | Whole | | Skimmed | UHT | Sterilised whole |
	Channel islands	Friesian			
water	06.4	07.0	91.3	87.8	87.8
total fat	4.9	3.9	0.1	3.9	3.9
(of which saturates)	3.1	2.5	0.06	2.5	2.5
protein	3.6	3.2	3.3	3.2	3.2
carbohydrate (lactose)	4.6	4.6	4.8	4.6	4.6
calcium / mg	131	115	120	115	115
vitamin A / µg	60	53	Trace	53	53
riboflavin / mg	0.19	0.17	0.18	0.17	0.17
vitamin D / µg	0.04	0.03	Trace	0.03	0.03
energy / kJ 100g^{-1}	320	276	140	276	276

BACKGROUND

Since the domestication of sheep, production of wool and later manufacture of wool products from sheep has been very important to many human communities, worldwide. In Britain, 200 years ago, there was a thriving wool industry and sheep were farmed for their wool rather than meat, but now for sheep farmers the income from wool has dwindled to less than 10% of their annual income, though it is a little higher for hill sheep farmers. The value of the fleece just about covers the cost of shearing. In Australia and South America, the position is reversed and wool remains the primary product of the sheep industry.

Wool production by sheep

Wool has special properties that, through the centuries, have made it a suitable and comfortable fibre for clothing. Wool normally holds about 16% of its weight as water and can absorb about as much again before it begins to feel wet. It actually gives out heat as it absorbs water so gives excellent protection against chilling. Even in a hot climate, sweat can be transmitted through woollen clothing. The air trapped amongst the soft fibres provides insulation and the feeling of comfort and warmth.

Wool is a form of hair, made up from **keratin**, the protein also found in horn, hooves and feathers. Like hair, the wool fibres grow from follicles in the skin. These follicles are of two types and produce different types of fibre. It is these fibres that determine the particular properties of the wool from different breeds. The nature of the **primary** and **secondary follicles** is illustrated in

Figure 4.32 and descriptions of the fibres are summarised in Table 4.10. The number (or density) of follicles contributes to the quality of wool, and this is determined during gestation and around birth. The fibres have scales on the surface and these help to hold the wool fibres together when being processed. The ratchet like edges mean the fibres can move in only one direction and, particularly when warm and wet, if the fibres are rubbed they move in different directions and then become tangled. This property leads to felting or shrinking in fabrics that have been woven. The waviness or **crimp** of wool is the result of kinks in the follicles. Crimping is influenced by the rate of growth – the wavelength is longer when the growth rate is faster. The sebaceous glands produce a waxy **sebum**, known as **lanolin**, which is valued as a base in ointments and some cosmetics. On living sheep, the sebum helps to keep the skin and wool from drying and also has a bacteriocidal action. The watery secretion of the sweat glands acts as a natural detergent.

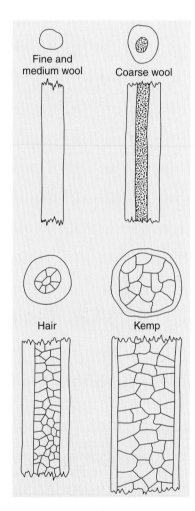

Why has the wool industry become less important in the textile economy and why do you think there has been a slight upturn in demand for wool in recent years?

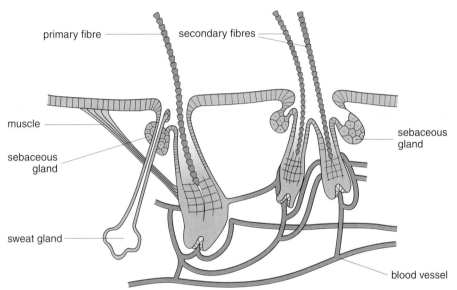

Figure 4.30 Vertical section through skin of sheep, showing wool fibres. Primary follicles have a sweat gland and erector muscle, secondary follicles are simpler.

Growth of wool is influenced by several factors, including the season, nutrition and the breed. In wild sheep, there is a seasonal cycle controlled by daylength, with most growth during the summer then less in winter followed by moulting in the spring. While most domestic sheep do not moult, there is a decline in growth during the winter (short-day) months. In the UK, 80% of the growth occurs between July and November. A good level of feeding is important for producing high quality wool. During periods of inadequate nutrition, the follicles

Figure 4.31 Range of different types of fibres in sheep. The diameter of the fibres can range from about 150 μm for kemps to 12 μm for fine wool. Another type of fibre, known as a heterotype, has a medulla like kemps during the summer, but in winter heterotypes lose their medulla and continue to grow like wool.

Table 4.10 *Summary of types of wool fibre, their properties and origin in primary or secondary follicles*

Follicle	Type of fibre	Description of fibre
primary	hairy fibres	long, thick; with medulla (hollow centre)
	kemps	short, thick; with medulla that nearly fills the centre so these fibres become brittle; pale in colour, stop growing in winter
secondary	wool fibres	short, fine, soft; no medulla

produce thin and weak fibres, and growth of wool is generally less during pregnancy and lactation, or during spells of disease. Poor nutrition before or soon after birth limits the number of follicles giving poorer wool quality.

The breed of the sheep probably makes the greatest contribution to fleece and wool quality. Since the early days of domestication, selection has been for sheep with fleeces that are white and woolly and which grow continuously allowing them to be shorn. There has also been progressive loss of the moulting character, and divergence into about four recognisable types. The fineness is a reflection of the number of secondary follicles - a high S/P ratio (secondary : primary) gives a finer wool.

- fine – as found only in the Merino breed – S/P ratio ~ 20:1
- semi-fine, shortwool (eg English Down breed) – S/P ratio ~ 5:1
- medium fibres (eg Longwoolled breed) – S/P ratio ~ 4:1
- hairy (eg Scottish Blackface) – most of the kemps have developed into long continuously growing hairs – S/P ratio ~ 3:1

For the sheep farmer, quality of wool is determined by a number of features:

- **fibre diameter** – fine wools are often preferred, though particularly outside the UK in hotter regions the sheep yield coarser fibres which are used in blankets or carpets. Diameter is measured in the laboratory in μm, though fineness is also expressed as a **Bradford Count**. Originally this was an estimate of the number of hanks, 510 m long, that can be spun from 450 g of wool. In practice, this measure is determined subjectively by the wool grader. Fineness varies in different parts of the fleece, the best coming from behind the head and shoulders and poorer quality towards the tail. Fibre diameters range from about 18 to 26 μm for the finest wools to 38 μm or more in the coarse wools. The corresponding Bradford Counts are around 80 for the finest to below 40 for the coarse wools.
- **staple length** – is simply the length of the longest fibres and may range from about 50 to 76 mm in shortwool breeds up to 300 to 500 mm in longwools. The longer staple lengths are required for carpets and weaving certain fabrics.
- **soundness** – is a measure of the tensile strength of the fibres. If easily broken, they are less suitable for further processing and this may be the result of poor nutrition during growth.
- **colour** – white wools are preferred because they can then be dyed to other colours. Some dark wools are used, particularly in traditional weaving, to make attractive patterns or for natural dark colours.
- **yield** – there is considerable variation between breeds, with longwools such as Devon and Teeswater yielding a fleece weight of 6 kg, whereas Swaledale and Herdwicks give a fleece of about 2 kg. The fleece should be as clean as possible, so usually dirt is removed, particularly around the tail, before shearing.

Shearing of sheep is generally done once a year. In the UK, this is best done in May or early June though will be a little later in the hill areas. Some shearing may be carried out in winter for sheep that are housed indoors. Shearing is a skilled task, and a good shearer will remove the whole fleece together without cutting the skin of the sheep. The fleece should be dry and kept clean. After removal the fleece is rolled in a particular way before being passed on for marketing.

Figure 4.32 Sheep being shorn

Chickens

The poultry industry

Development of breeds for eggs and meat

When first domesticated, chickens were probably used for cockfighting and later may have assumed a religious significance in some communities. It was probably considerably later that they became important as a source of food, providing eggs and meat. There is evidence of domesticated chickens in south east Asia between 4000 and 5000 years ago; 2500 years ago, the Greeks kept chickens for cockfighting; the Romans built poultry houses, understood the need for hygienic conditions and kept chickens both for religious ceremonies and for food. Any selection done by primitive people was probably for those chickens showing prominent spurs, or perhaps the largest eggs, and these were retained as parents for the next generation. Considerable interest in controlled selective breeding developed in the latter part of the 19th century, resulting in the rapid evolution of a range of classes, breeds and varieties of poultry. The focus was on fancy, decorative and exotic breeds, for competitive showing. Little attention was paid by these breeders to productivity in terms of eggs or meat. Show breeding became an end in itself, isolated from those with commercial concerns, and has persisted until today as a specialist hobby.

The red jungle fowl, *Gallus gallus*, is the ancestor of an array of modern forms of the domestic chicken (*Gallus domesticus*). Today, if you walk from a village in Thailand into the jungle, or in many other locations through south east Asia, you might be rewarded with glimpses of the colourful wild jungle fowl flying through the trees or certainly hear the familiar noise of its call. The untrained eye may be unable to distinguish it from the domesticated cockerel strutting in the village street or even bantams scattered today in backyards throughout Britain.

Figure 5.1 In a Thai village – a domesticated chicken closely resembles wild jungle fowl

Figure 5.2 Showbred chickens (White Yokohamas)

Development of breeds for the production of eggs and meat also started during the 19th century, particularly in the United States and England. Programmes of breeding and selection concentrated on relatively few characteristics, notably mass production of large, white-shelled or brown-shelled eggs and of chickens raised for meat (known as broilers). White-shelled eggs come almost exclusively from the variety known as White Leghorns; brown-shelled eggs come from other varieties, such as Rhode Island Reds, New Hampshire or Barred Plymouth Rock. Most of the chickens used for broiler meat come from crosses between White Cornish and White Plymouth Rock. By the 1980s, a few very large multinational breeding companies had become responsible for providing and distributing stock on a world-wide basis. This has led to concern that the genetic pool has been reduced to a

CHICKENS

QUESTION

The choice of colour of eggs is largely market-led – white eggs are preferred in the United States. How would you find out which colour is the more popular in Britain? What do you think determines people's preference?

dangerously low level. It may be that show breeds, or those chickens which have persisted in village economies (or even the wild jungle fowl), may assume renewed importance in scientific and commercial fields, providing a reservoir of genetic diversity, a bank of genes with the potential for introducing new or desired characteristics into modern poultry stocks.

The modern poultry industry

The commercial farmer aims to maximise productivity in terms of yield of eggs from laying hens and meat from broilers. The swing to **intensive systems** occurred during the 1960s and 70s. By the 1980s, over 80 per cent of laying hens in Britain were managed in battery systems and over the same period, the **size of flock** increased. The development of a high level of **automation** has contributed to the rapid increase in intensive systems, though it involves considerable capital outlay in terms of housing and other equipment.

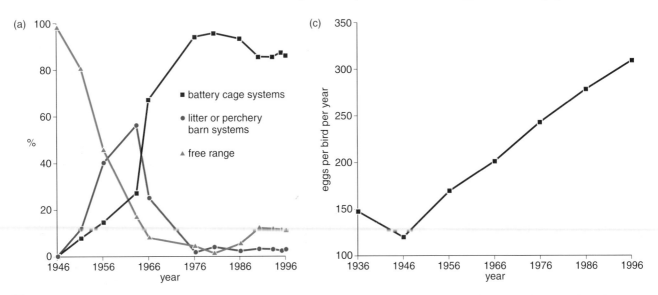

(b)
Size of flock (in Britain)

laying hens

1957 85% *under* 1000 birds

1971 85% *over* 1000 birds

1986 50% *over* 50 000 birds

broilers

1991 50% *over* 100 000 birds

*Figure 5.3 Development of the poultry industry – intensification and increased productivity: (a) systems of management – comparing battery cages, litter or perchery (barn) and free range systems; (b) size of flock; (c) productivity as **number of eggs per bird (layers)**. Improvement in productivity in **broilers** can be seen by the time to reach 1.8 kg live weight. In 1957 they took 68 days whereas in 1987 this was only 38 days*

Increased intensification has been accompanied by improved **performance** per bird. For laying hens this is seen as an increase in the number of eggs over a period of time and for broilers as a reduction of the time they take to reach the desired weight. The higher productivity achieved can be attributed to several factors:

- genetic improvement through selection of higher yielding strains
- development of improved breeding techniques
- understanding of nutritional requirements
- better management and control of the environment, particularly housing
- measures to control disease.

The bulk of intensively farmed chickens are reared in houses, under controlled environmental conditions, though there are outdoor systems, using yards or 'free range', and also various types of semi-intensive, small-scale operations.

Recent years in Britain have shown increasing demand for free range eggs and poultry meat, but mass production or 'factory farming' still predominates. Small-scale production is, however, practised in rural economies with subsistence farming, especially in developing countries. Here it is particularly valuable because of the low initial capital input and ability of chickens to scavenge and use local feeds.

Behaviour and its implications for the poultry industry

Certain behaviour patterns seen in wild jungle fowl can be recognised in small domesticated groups of chickens. Expression of such behaviour has implications for birds kept under intensive conditions, because of the effect on productivity and the need to consider welfare. The red jungle fowl is gregarious, often coming out of dense forest into clearings in the morning and evening to feed. Its diet is varied, including insects, worms, plant seeds and other vegetation, scratched and pecked from the surface soil. It remains wary, ready to dart back to cover, roosting on branches off the ground. Familiar behaviour patterns in domesticated chickens include 'dusting', persistent scratching of the ground and showy courtship displays by the cocks. During the breeding season, a single cock maintains a territory with about five hens. Eggs are laid in a nest on the ground in clutches of three or four, usually between March and May. The hen sits on the eggs during the 21 days of incubation leading to the hatching of the eggs, then shows protective behaviour towards the chicks before they become independent after a few weeks.

Within a group of chickens, a social order develops, known as the **pecking order** (Figure 5.4). Evolution of a pecking order may be seen as a mechanism for sharing out space, allowing distribution of the birds within the space without resorting to fighting. Once the territory is established, there is no need for pecking. If space becomes limiting, aggressiveness may result. Communication between birds is by a range of visual postures which reinforce the social status, including physical pecking. When new birds are introduced into a flock, there is a period during which their position in the pecking order is established. In a confined space, such as a yard or hen-house, chickens still try to establish a territory within which they move. Straying beyond the arbitrarily defined limits may result in an increasing amount of pecking from other birds. In a large flock, subgroups become defined, within which the hierarchies are established. With laying hens in a confined space, those lower in the pecking order tend to produce fewer eggs, but this effect is avoided with laying hens kept in battery cages. At low light intensity, as occurs towards dusk, demand for space required is reduced, hence the calming effect of low light on birds.

Chickens also communicate by a range of calls, described variously as cackles, clucks, squawks and so on, each with a different meaning. These include, for example, calls which indicate alarm, or those associated with finding food, with courtship and with egg laying. Broody hens have a series which allows elaborate communication with their chicks. On an individual basis, chickens show preference for social association (or companionship) rather than being kept in isolation. They also scratch the ground (or litter in artificial systems)

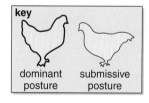
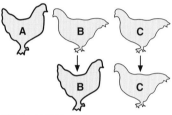

simple pecking order

key

dominant posture   submissive posture

Figure 5.4 Simple pecking order in chickens. The dominant chicken is shown with a heavy outline; those in a lower pecking order are shown with a thin line. In the pecking order for a group of hens, the dominant hen A can peck hen B and hen C, but neither can peck her. This hierarchy means that, for a group of hens, the dominant hen can peck all below but none peck her. Similarly the second in the hierarchy can peck all below but is submissive to the dominant hen and so on. Separate pecking orders may develop for males and females kept together.

and bathe in the dust as a means of of cleaning themselves. Deprivation of both space and suitable ground cover, as experienced in intensive systems, may lead to poor health, loss of feathers and extremes of behaviour.

Reproduction and breeding

Male reproductive system and mating

The reproductive system of male birds consists of a pair of yellow **testes** (which remain within the body cavity), the **epididymis** and **vas deferens**. A courtship display is followed by the mating process – while the hen squats, a cock bird grasps her by the skin at the back of the neck then mounts her. Sperm held in the vas deferens are ejaculated in semen into the entrance of the oviduct. As the hen rises to a standing position, the spermatozoa move up the oviduct towards the infundibulum, where fertilisation normally occurs.

In a breeding flock, seven to ten males are allowed for every 100 hens. Artificial insemination (AI) is used widely in the turkey-breeding industry where the extreme development of a heavy breast for meat production has resulted in birds that are almost incapable of mating naturally. In chickens, use of AI is limited, due more to the labour costs than to lack of technological development. There is potential for the future genetic improvement of breeding stock of chickens, particularly broilers, with the use of AI.

Female reproductive system

Two ovaries are present in the developing embryo but only the **left ovary** and **oviduct** become functional. This ovary may contain 2000 to 12 000 or more ova of different sizes, but in modern poultry practices, probably only 200 to 300 of these are ovulated. The **ova** develop in sequence and are released in order of their relative size, the largest first. In a laying hen the largest are several millimetres in diameter and are visible to the naked eye in a dissected ovary. There are likely to be five or six large, yellow developing egg yolks (follicles) and a large number of very small, white follicles which represent immature undeveloped yolks. The Graafian follicle, filled with yolk, contains the **ovum** together with its surrounding membranes. The follicular membrane attaches the follicle to the ovary and is well supplied with blood vessels. At ovulation, the whole structure is released from the ovary and develops into the future yolk of the egg. When released, the yolk has reached its full size which is directly related to the size of the egg when laid. The reproductive cycle is influenced by **light**. Longer **daylength** promotes the laying of eggs whereas shorter daylength results in a decrease in production. Several hormones are involved in the reproductive cycle and their functions are summarised in Table 5.1.

Ovulation to egg-laying

During the passage of the ovum along the **oviduct**, protein and carbohydrate components of the egg are added as well as the materials of the shell. At

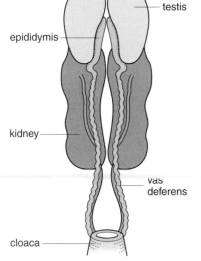

Figure 5.5 Male reproductive system of a chicken

Labels: testis, epididymis, kidney, vas deferens, cloaca

ovulation, the yolk mass containing the ovum is released and is directed into the **infundibulum** then passes into the oviduct. The infundibulum also provides a short-term storage site for sperm. Fertilisation may occur either

Figure 5.6 The relative sizes in the follicular hierarchy of successive ova in a hen's ovary. The ova develop in sequence, with the largest being released first

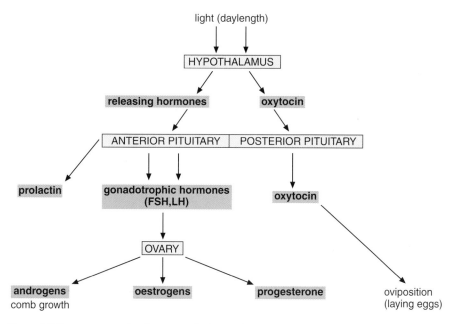

Figure 5.7 Hormones involved in control of reproduction

Table 5.1 *Summary of roles of different hormones involved in reproduction*

Hormone	Role
prolactin	• broodiness
follicle stimulating hormone (FSH)	• initiates development of the ovarian follicle
luteinising hormone (LH)	• stimulates ovulation of the mature yolk from the ovary. Daylength influences the timing mechanism which controls release of LH from the pituitary gland
oestrogen	• female plumage • mating and nesting behaviour • development of the oviduct, which is small and undeveloped in immature females • promotes formation of medullary bone tissue in the narrow cavity of the femur and some other bones – this acts as a store for calcium, with heavy demands at the time of egg-laying for formation of the shell • increase in blood levels of calcium, protein, fat and vitamins, needed for egg formation • softening and spreading of the pubic bones, enlarging of of the vent, in preparation for the actual laying of the egg (oviposition)
progesterone	• stimulates hormone releasing factors of the hypothalamus to release LH from the anterior pituitary • ovulation • albumen formation
oxytocin	• expulsion of the egg after it has passed through the oviduct
male sex hormone	• contributes to bright red of the comb and wattles in laying hen

here or as the ovum enters the magnum. The **magnum** is white, with thick walls and a relatively large diameter. It has prominent mucus-secreting cells in the walls, including a large number of ciliated cells covered in albumen. In the magnum, the albumen (egg white) is secreted in layers around the yolk. The albumen is uniform as it is secreted, but becomes differentiated into thick and thin layers because of twisting movements of the ovum as it passes along the oviduct, and water is also added. Peristaltic movements carry the albumen-coated ovum into the isthmus, which is narrower than the magnum. The two shell membranes are produced in the isthmus. The membranes consist of the protein keratin, the outer being thicker than the inner. They are partially permeable, allowing passage of water and salts. The **tubular shell gland** is involved in the transfer of calcium salts to the shell membranes. The uterus, also called the **shell gland pouch**, is the region where fluid is added. Up to here the egg has been flaccid, but the fluid causes it to swell into its characteristic shape. The bulk of the hard shell material is added here, in the form of calcium salts deposited on the shell membrane. Pigment may be added while still in the uterus. The colour of brown eggs is due to porphyrin pigments. The **vagina** makes no contribution to shell formation but contains **sperm host glands**, responsible for providing nutrients which enable the sperm to remain viable before being carried up the oviduct by cilia and antiperistaltic movements. Within the vagina some sperm can be stored and remain viable. This allows the eggs to be fertilised on successive days over a period of time, without the need for further mating. **Muscles** in the vaginal wall relax when the egg is to be laid, allowing it to pass quickly from the uterus to the cloaca. These muscles are under voluntary control so the hen can delay laying the egg when conditions are unsuitable. The completed egg leaves the oviduct through the **cloaca**.

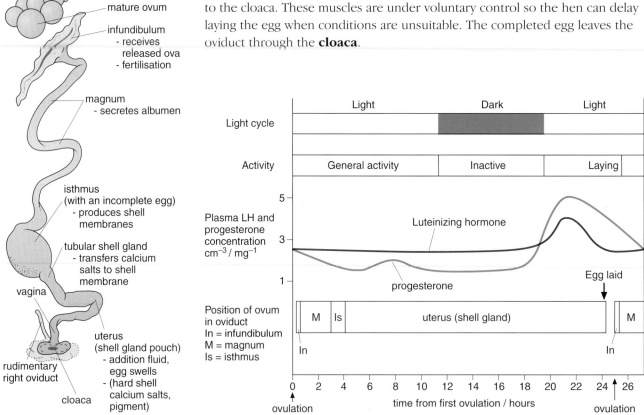

Figure 5.8 (a) oviduct of a hen and roles of each region during the laying of an egg; (b) laying an egg – events and activities occurring over a 24 hour period in a hen.

The whole process of laying an egg, from the time of ovulation to the deposition of the egg in its hardened shell (**oviposition**), takes about 28 hours, with about 20 hours being spent in the uterus. Ovulation usually occurs sometime after midnight and normally most eggs are laid only during daylight. Laying occurs progressively later each day over a period of time until darkness intervenes, then, after a gap of a day or so, a fresh **sequence** starts. This is equivalent to a clutch in wild hens. In highly productive hens, laying may continue for more days without a gap and the period of egg formation may be shortened to about 24 hours. In a flock of laying hens, the gap in sequences of different hens should smooth out, allowing reasonably steady production. A domesticated hen can continue to lay eggs without any need for mating and fertilisation. In commercial flocks of laying hens the eggs are infertile unless required for breeding.

Structure of the egg

Eggs vary in size. Those laid earlier in a sequence are slightly larger than the later ones. Eggs from younger birds, starting to lay at the age of 20 weeks, are smaller than those from older birds, say about 50 weeks. An egg is irregularly ovoid in shape (with one end more pointed and the other blunt), on average between 5 cm and 6 cm in length.

The familiar outer **shell** has important biological functions. The hard shell consists of inner and outer shell membranes enclosing an organic matrix of a glycoprotein into which salts are deposited. This hard shell is 98 per cent calcium carbonate with some magnesium, phosphate and citrate. Nearly 2 g of calcium is deposited in each eggshell, making very heavy demands on calcium in the diet. There may be as many as 8000 pores in the shell of an egg, arranged mainly above the equator towards the larger end. The pores have a diameter of about 20 μm and they provide a means for gas exchange. They are partly closed with a cuticle, which helps reduce moisture loss and prevent entry of bacteria. The outer, thin, dense layer of the shell is mainly responsible for preventing entry of bacteria whereas the inner layer is more open and granular, providing strength and rigidity. This layer also acts as a source of calcium for the growing embryo for calcification of its bones.

Modern systems of egg production, such as high rates of lay, use of battery cages and mechanical methods for collecting eggs, tend to increase the likelihood of physical damage to eggs. Short cycles of light alternating with dark improve the quality of shells, probably in response to the level of calcium in the intestine. Mechanical washing of eggs for sale removes much of the cuticle, so eggs are sprayed with an oily mist to replace its protective properties.

Internally, the bulk of the egg is made up of **albumen** (egg white), which is a complex mixture of about 40 different types of proteins. Albumen has a nutritional function for the developing embryo, but it also provides a source of water, acts as a bactericide, protects the yolk against mechanical injury and presents a surface for the deposition of the shell membranes. Within the albumen, alternating layers of thick and thin white can be distinguished, with

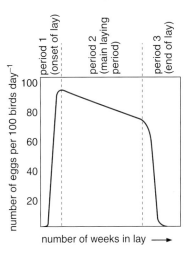

Figure 5.9 Egg production in a flock of laying hens

- *an individual hen of a laying strain comes into lay at an age of about 18 weeks and could reach a maximum of about 310 eggs in a year;*
- *eggs are laid in sequences, with a pause day when no egg is laid. In a flock, these pauses tend to smooth out so that a steady production is achieved;*
- *Period 1 is quite short and represents the time from the first eggs being laid until nearly all the birds in the flock are laying;*
- *in Period 2 the decline is due mainly to lengthening time for formation of the egg;*
- *Period 3 shows a rapid decline. The end of egg laying can be caused by broody behaviour, moulting, changes in nutrient intake or decreasing daylengths.*

CHICKENS

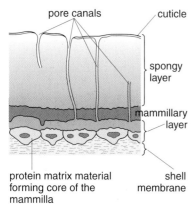

Figure 5.10 Section through the shell of a hen's egg. Calcium salts are laid down in the spongy layer and provide the hardness. The pores allow gas exchange and the cuticle reduces moisture loss and prevents entry of bacteria

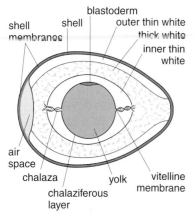

Figure 5.11 Internal structure of a fertilised hen's egg

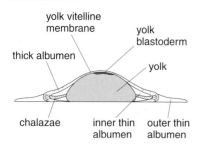

Figure 5.12 Cross section of broken out egg

Table 5.2 *For sales, eggs are first classified by quality then graded according to weight. Class A eggs have a clean cuticle and undamaged shell and are used as 'fresh' eggs. Internally the air space is less than 6 mm the albumen clear and the yolk visible only as a shadow when viewed over a light (candling). Class B eggs, used as second quality eggs or for processing, may have a dirty or damaged cuticle but the shell is undamaged. Class C eggs are used only in processing, and may be dirty with cracked shells, the air space larger than 9 mm and on candling the yolk is distinct. The table shows European Union grades and sizes for marketing of Class A and Class B eggs.*

Grade	Mass of egg / g	Grade	Mass of egg / g
0	75.0 or more	4	55.0 to 59.9
1	70.0 to 74.9	5	50.0 to 54.9
2	65.0 to 69.9	6	45.0 to 49.9
3	60.0 to 64.9	7	less than 44.9

the thick white making up about half of the total. The chalazae are coils of twisted mucoid fibres, derived from the albumen. They form a link between the very thin chalaziferous layer of dense white adjacent to the yolk and the main region of thick white. The turgidity of the albumen is important for providing support for the shell membranes. Watery albumen, say as a result of disease, leads to poor shell deposition. Eggs lose water from the albumen through the shell during storage, with a corresponding increase in the air cell. The size of the air cell can give a useful indicator of the length of storage time. The viscosity of albumen also decreases during storage, and this can be seen by the way the white of an old egg spreads when it is broken open.

The yolk is held within the albumen by the chalazae. In a fertilised egg, the germinal disc or blastoderm, which has the potential to develop into the embryo, can be seen as a whitish disc, 4 mm in diameter. It floats upright at the top of the yolk, held in position by the latebra (stalk) within the yolk. The yolk consists mainly of phospholipids, synthesised in the liver, and its yellow colour is due to carotenoid pigments. Occasional double-yolked eggs result from two yolks moving through the oviduct at the same time, usually because two ova are ovulated together. Blood spots sometimes appear on the yolk, due to irregular tearing of the follicle at ovulation. Such eggs are less acceptable for consumption. When the egg is laid, the surrounding temperature is cooler than the body temperature of the hen, so the egg contents contract, leaving the air cell at the blunt end, between the two shell membranes. The colour of the yolk is due to yellow pigments known as xanthophylls and can, to some extent, be influenced by diet. Grass and maize, for example, are rich in xanthophylls. Some poultry feeds need to include supplementary pigments to ensure the desired yolk colour is achieved for market demands.

Development and hatching of fertilised eggs

A fertilised egg takes 21 days to develop until the chick hatches. The temperature must be maintained at about 38 °C with a humidity around 90 per cent. In the natural state, the hen lays her clutch of eggs then becomes 'broody'. This brings about a change in her behaviour pattern which ensures that she sits on the nest and maintains the required conditions for incubating through to hatching. Broodiness is part of the natural reproductive cycle which starts with courtship and mating, followed by nest-building and laying of the clutch of eggs, and is closely linked with events in the ovary. Secretion of prolactin from the anterior pituitary gland, increases during incubation and decreases during the parental phase, after the chicks have hatched. The first few eggs laid by the hen can be left for a few days before they start to develop so that the clutch of eggs can be completed. Sitting must then be more or less continuous. A natural broody hen turns her eggs several times a day which keeps the embryo and yolk free from the membranes. She maintains the temperature with her body warmth and the feathers provide insulation and help maintain the required humidity.

Eggs required commercially for development into chicks are incubated in hatcheries under controlled artificial conditions. Several thousand years ago huge incubators were known in China and in Egypt. In modern poultry practices, fertilised eggs from laying hens are collected several times daily and can be stored for up to a week at 15 °C and 75 per cent relative humidity. If kept longer than this, they should be turned. In a commercial incubator, eggs are placed with the large end uppermost as this position allows the embryo to develop near the air cell, and also makes it easier for the chick to emerge from the egg. At first, the temperature is maintained at 37.8 °C with adequate ventilation to ensure even temperatures and allow for gaseous exchange. Three days before the eggs are due to hatch they are transferred to a different incubator with higher humidity and a temperature of 36.9 °C. During incubation, short periods of cooling can be tolerated, as would occur in a naturally sitting hen, but embryos are easily damaged by cooling during the last two days. This is the time when a hen sits 'tightly' on her nest. The embryo itself generates heat so adequate ventilation is needed to ensure excess heat is lost. The high humidity prevents rapid loss of moisture from the egg which would lead to drying out of the embryo. Throughout the incubation, turning is done mechanically by automatic rotation of the trays of eggs every few hours. Levels of gases are also monitored, particularly carbon dioxide, since hatchability is noticeably decreased if the level of carbon dioxide rises to about 2 per cent. Reduced oxygen also decreases the percentage that hatch successfully. Newly hatched chicks are removed. Eggs which do not hatch may include incompletely developed living embryos and must be destroyed rapidly according to recognised codes of practice. A very high standard of hygiene is required in the hatchery to minimise diseases, including those passed from hen to chick through the egg.

Internal changes take place within the egg and lead to the development of the chick. The ovum is fertilised within 15 minutes of ovulation and divisions of the fertilised ovum begin during its passage through the oviduct. The fertilised egg (zygote) divides by mitosis, first into two cells, then these each divide into four, then double to eight and then to sixteen and so on. The first division takes place about 5 hours after fertilisation, as the egg enters the isthmus; subsequent divisions take place so that it has reached the 256-cell stage about 9 hours after fertilisation.

Cell division continues and development resumes when incubation starts. Four membranes develop outside the embryo, forming sacs known as the yolk sac,

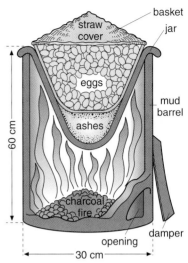

Figure 5.13 Ancient Chinese incubator – compare this with natural hatching of eggs by a broody hen and a modern incubator. How are the essential conditions achieved in the ancient Chinese model and what do you think the drawbacks would have been? Why do you think they used these incubators?

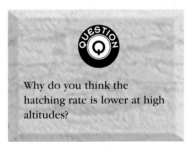

Why do you think the hatching rate is lower at high altitudes?

Identify features in a modern artificial incubator which simulate conditions achieved during natural incubation by a sitting hen.

Figure 5.14 Partially developed embryo, as seen on yolk broken out of egg

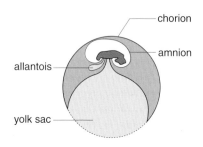

Figure 5.15 Embryo developing on top of the yolk – about day 4

Figure 5.16 Newly-hatched chicks

Why do you think chicks are sexed? Think about future laying flocks as well as broiler chickens.

amnion, chorion and allantois. The **yolk sac** is well supplied with blood vessels which enable nutrients to be carried from the yolk into the embryo. The **amnion** and **chorion** fold around the whole embryo and contain a fluid, part of which is derived from the albumen. The fluid bathes the embryo, allowing it to float freely during development, giving protection from dehydration and mechanical shocks, and insulation from temperature changes. The **allantois** is well supplied with blood capillaries and provides the respiratory surface for exchange of gases until the lungs take over the day before hatching. Waste nitrogenous material is passed into the fluid in the allantois.

During development, the embryo derives its nutrients mainly from the yolk sac. Just before hatching, the remains of the yolk sac are drawn into the body of the chick and this can sustain the chick for two or three days after hatching. Calcium required for the skeleton is absorbed from the shell, which becomes weaker thus helping the hatching process. As the embryo develops, the air cell increases in size. It is probably the composition of the gases in the air cell which initiates the chick's urge to emerge from the shell. Oxygen in the air cell may fall to about 16 per cent and carbon dioxide may rise as high as 4 per cent. The embryo starts to breathe air through the lungs on the 20th day. When ready to emerge from the shell after 21 days, the chick repeatedly jerks its head, using its beak to break first through the allantois, then through the membrane separating this from the air space. Over a period of several hours, the shell fractures or 'pips', allowing the chick to escape. The newly hatched chick is wet and exhausted, but within a few hours it dries out to a fluffy chick, able to move around actively.

In commercial hatcheries, newly hatched day-old chicks are usually sexed. This can be done by examination of the vent, but the practice requires skill, is slow and may be stressful to the chick. Alternative methods make use of sex-linked characters which are detectable at this age, for example feather-sexing (day-old females having longer wing feathers than males) or colour-mating producing males with white or yellow-coloured down and females with buff or red. Within the first 24 hours, chicks can be transported considerable distances, provided they are packed in suitable containers with reasonable temperatures. During this time no outside food supply is required as the chicks use up the remains of the yolk sac. Large-scale hatcheries have become a specialised part of the poultry industry, supplying day-old chicks to intensive units either to become layers to produce eggs or to be grown into broilers for meat.

Growth and development of the chicks

A broody hen will naturally continue to protect her chicks, jealously guarding them against intruders, keeping them warm by spreading her feathers over them and leading them to food. Over a period of a few weeks they gradually gain their independence. In commercial houses, the brooding requirements are artificially simulated. Growing chicks, reared from day-olds, need to be kept warm. At first they require a temperature of 35 °C, then over a period of three weeks this is reduced in three stages to 24 °C. The heat should be well distributed in the houses, allowing the chicks to space out and have freedom to move (see Figure 5.17). They should have free access to crumbs of food, which may be spread on belts on the floor with fresh water continually available at first from pipes with nipples and later from containers designed to avoid spillage.

Figure 5.17 Young chicks on floor of rearing shed, with adequate supplies of food, water from nipples on the pipe, and warmth from lamps

Almost by default the shell provides packaging material for the egg. What part does the shell play in the marketing of the product? How does it help keep the contents fresh and what are its limitations?

Egg production

The main concern of the egg producer is to maintain a supply of eggs that matches market demands. Eggs may be marketed directly for consumption or fertilised eggs are hatched and reared into broilers for meat or kept to provide replacement laying hens for eggs.

The modern hen has a very high rate of egg-laying, averaging about 250 per year. This has been achieved by selection of productive strains and control over the environment. Extension of each sequence of egg-laying is achieved mainly by artificial manipulation of the lighting regime, stimulating response to the photoperiod. Day-old chicks are usually started with 23 hours of daylight, which allows them to find food and water. After 1 or 2 days this is reduced to an 8-hour daylength, sometimes in steps over a period of a few weeks. After 18 weeks, daylength is increased at the rate of about 20 minutes per day, to reach the ideal daylength of 17 hours which is suitable for maintaining hens in lay. Where natural daylight is being used, the daylength can be extended during certain seasons by artificial light morning and evening. Young hens (pullets) usually start to lay at the age of 20 weeks. A hen tends to lay for a period, then undergoes a period of moulting when egg production ceases. Moulting can be induced artificially by altering lighting and feeding regimes. On recovery from moulting, the hen comes into lay again.

There are several systems, from free range to highly intensive, for keeping laying hens. The differences relate mainly to the space available, hence the degree of movement possible and association with other birds (see Table 5.3). In **free-range systems**, birds must have access to open-air runs and be provided with housing for the night and for laying of eggs. The land should be covered with suitable vegetation (usually grass), with adequate supplies of fresh water and some shelter. In free-range systems the land can suffer from over-use and become 'fowl sick', with a build up of organisms causing disease. The vegetation deteriorates, partly because of the pecking, dust bathing and other activities of the birds but also because of the excess of fresh poultry manure which is rich in nitrogen and rather acid. These difficulties can be overcome by rotating the area of land being occupied, using portable houses, and control of the stocking density. Protection against predators such as foxes and even dogs and cats should be provided. In **semi-intensive systems**, outside access is available in wire enclosures, used in rotation, with housing that is more likely to be permanent in its position. **Straw yards** are open to the weather but have protection from a roof. Plenty of litter, usually as chopped straw, is provided, together with nest boxes and perches.

For indoor systems, the **deep litter system** gives the birds some freedom of movement, and the floor is covered with a deep layer of wood shavings, peat or other material. Perches and nest boxes are provided. This system is useful for breeding flocks as males can be included allowing mating to occur. Disadvantages are that eggs may be laid on the floor, becoming dirty and needing special collecting. Another difficulty is that pecking orders become established and, in the relatively confined space, birds low in the pecking order may become deprived of food or suffer physical injuries. The perches should be arranged so that most of the droppings fall into a pit or limited area which can then be removed, keeping the rest of the litter reasonably dry.

The **battery cage system** accounts for over 90 per cent of the eggs produced commercially in the UK. A high degree of automation is used at all stages to increase efficiency and lower the costs of production. Attempts have been made to devise systems allowing more freedom of movement for the birds with a 'get away' area offering a nest box and perch. The **aviary** and **perchery** systems give different arrangements for stacking of the cages. Cages may hold up to six birds and they may be arranged in different ways to allow adequate access and removal of droppings. Food is generally provided in troughs by a moving chain feeder and water is continually available from nipples or drinking troughs. A sloping floor allows eggs, as they are laid, to roll onto conveyor belts which carry them straight to the cleaning, grading and packing area. Similarly, provision is made to remove droppings, either from a moving belt or

Table 5.3 *Outdoor and indoor systems for keeping laying hens. With increasing intensification, consideration must be given to the welfare of the hens in relation to their natural behaviour patterns*

System	Space
Outdoor systems	
• free range	400 birds per ha
• semi-intensive	1000 birds per ha
• straw yard	3 birds per m^2
Indoor systems	
• deep litter	7 to 10 birds per m^2
• aviary, perchery, multi-tier	20 birds per m^2 of floor space
• laying cage	450 to 750 cm^2 per bird

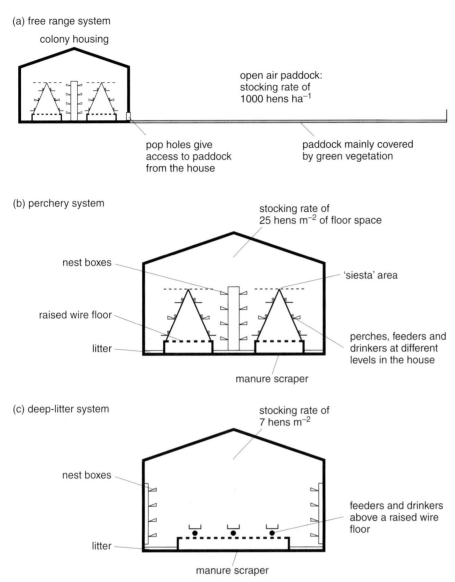

(a) free range system

colony housing

open air paddock:
stocking rate of
1000 hens ha^{-1}

pop holes give
access to paddock
from the house

paddock mainly covered
by green vegetation

(b) perchery system

stocking rate of
25 hens m^{-2} of floor space

nest boxes

'siesta' area

raised wire floor

litter

perches, feeders and
drinkers at different
levels in the house

manure scraper

(c) deep-litter system

stocking rate of
7 hens m^{-2}

nest boxes

litter

feeders and drinkers
above a raised wire
floor

manure scraper

Figure 5.18 Types of housing used for laying hens (a) free range; (b) perchery system;
(c) deep-litter

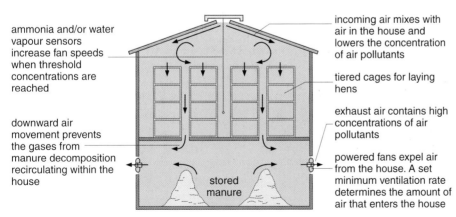

ammonia and/or water
vapour sensors
increase fan speeds
when threshold
concentrations are
reached

downward air
movement prevents
the gases from
manure decomposition
recirculating within the
house

incoming air mixes with
air in the house and
lowers the concentration
of air pollutants

tiered cages for laying
hens

exhaust air contains high
concentrations of air
pollutants

stored
manure

powered fans expel air
from the house. A set
minimum ventilation rate
determines the amount of
air that enters the house

Figure 5.19 Management and control of the environment inside houses for caged laying birds

of the environment in terms of lighting (intensity and duration), temperature, humidity and ventilation. Criticism of and public concern relating to the battery system on welfare grounds comes because the birds are restricted to an extreme in their movement. They are unable to take exercise, dust bathe or form natural social associations with other birds; they lack the privacy characteristically shown by birds during egg laying, and may suffer from breast blisters, lameness and loss of feathers. Lack of feathers removes the natural insulation provided, so temperatures in the houses must be higher to compensate. Arguments in favour of battery systems are that the birds are in a controlled environment and are not exposed to climatic extremes or predators and, under good management, disease risk is low. From the commercial point of view, and the consumer, the low cost and scope for efficient management is attractive.

Broilers for meat

In the broiler industry, systems are geared to producing birds of a desired weight in a predetermined time (say 2 kg in 41 days), and keeping costs as low as possible. A number of units work in phase to provide a continuous supply of meat, based on predictions in a fluctuating market. Often processing and 'value added' treatments are included under the same management. This helps to ensure smooth and efficient flow from day-old chicks to the final marketable products. The largest single annual cost is that of food and strict precautions must be taken to minimise losses due to disease.

Birds are typically housed in single storey wooden buildings, grouped on one site, isolated from other nearby sites to reduce the risk of transfer of disease. The house provides a controlled environment for the birds from their arrival from the hatchery as 1-day-old chicks until they reach the desired weight in about 6 weeks. Ventilation is achieved by air inlets in the sides of the house and vents in the roof. Circulation may be enhanced by electrical fans. In hot weather, air movement may be considerably increased or even reversed to assist with cooling. Ventilation is necessary to remove moisture released from respiration thus avoiding excessive condensation. This would lead to release of ammonia from waste products accumulated in the litter. Houses should be well insulated to avoid temperature extremes, both hot and cold, though artificial heat is used when required. Materials used for litter include straw, wood shavings or straw pellets. Litter is important both to provide insulation and to absorb the droppings from the chickens. A relatively low level of lighting is normally used throughout, though sometimes chickens are kept in darkness for a short period of, say, 2 hours in 24. Because of the dependence on electricity, emergency generators are advisable in case of power failure. In closed houses, a maximum stocking density of 20 birds per m^2 is allowed (see Table 5.3).

A continuous supply of clean, fresh water is essential, and the rate of consumption of water gives a useful measure of growth rate. Water may be provided in troughs or through pipes with nipples along which the water flows under controlled pressure. Generally water containers are raised higher above the floor level as the birds grow. Food is supplied at first (for day-old chicks)

on strips on the floor, then from troughs or moving belts. The physical form and nutritional composition of the rations change as the birds grow. For about the first 11 days, a 'starter' feed as crumbs is given to the chicks, followed by 'grower' pellets to between 3 and 4 weeks and a 'finisher' up to the time of slaughter. Some of the protein in the food originates from white fishmeal and different types of soya preparations. The feed also includes a cereal fraction, originating from wheat, maize or barley. Levels of specified mineral salts are also controlled.

After the required number of days (usually between 42 and 56) most of the birds reach the desired weight. The whole flock is removed from the house, transported to the processing unit where in a matter of minutes the birds are stunned, killed, plucked, eviscerated and prepared for the next stage, which may be for immediate sale, for freezing or further value-added processing. The houses are completely cleared out, cleaned and disinfected, ready for the next arrival of day-old chicks a few days later.

In recent years there have been concerns over cases of food poisoning caused by the bacterium *Salmonella enteritidis* and links with eggs, egg products and inadequately cooked poultry meat. Broken eggs provide an ideal source of nutrients for growth of *Salmonella* or any other bacteria, and similarly use of eggs with shells that are cracked or contaminated with excreta may lead to infection. Increases in cases of food poisoning during the 1980s led to the identification of a new strain of *S. enteritidis*, known as PT4 (phage type 4), which can infect ovaries of hens and persist in some laid eggs, even when undamaged. These bacteria usually occur on the surface of the yolk membrane and often no further multiplication takes place so the risk if infection is very low. If the bacteria reach the yolk or if cooking is insufficient to destroy the *Salmonella*, there is increased risk of a food poisoning outbreak.

In the egg, what part does the albumen play in minimising multiplication of Salmonella? How long after infection do the symptoms of Salmonella food poisoning appear?

The Farm Animal Welfare Council considers that there are five basic 'freedoms' that farm animals should be allowed. These are:
• freedom from hunger and thirst
• freedom from discomfort
• freedom from pain, injury or disease
• freedom to express normal behaviour
• freedom from fear and distress
How far do intensive systems of poultry production allow these freedoms to be achieved? Compare free range and indoor systems and think also about any benefits chickens may have gained from domestication. Then list some of the environmental consequences of intensive production – benefits and disadvantages.

6 Fish and fish farming

Fish farming was probably practised about 4000 years ago in China, where carp was the first species known to be domesticated. There are references to ponds stocked with fish in the Bible and in writings from ancient Egypt. In the monasteries of medieval Europe fish were kept to provide fresh fish on days when eating meat was prohibited for religious reasons. Marine fish farming started in Indonesia about 600 years ago, and has shown considerable expansion since the 1960s, particularly in Japan and other parts of Asia. In Europe, development of trout and salmon culture on a commercial scale began in the late 19th century, grew in importance during the first half of the 20th century and has shown a rapid escalation over the last 20 or 30 years. This is partly a response to declining stocks of wild fish.

The fish farming industry

Compared with mammals or birds, the degree of domestication in fish is far less and there is relatively little difference between wild fish and those bred and reared in captivity. Fish farmers are now actively looking for wild species showing characters with the potential for adapting to cultivation in an intensive environment, and there is scope for future development of controlled breeding for genetic change. There has been some artificial selection and change in domesticated carp, a species which has remained important in China and other Asian countries. While there is considerable potential for selective breeding to choose economically desirable characteristics (such as growth rate), generally the development of breeds of fish, as seen in cattle, sheep or chickens, is in its infancy.

In the mid-1990s **aquaculture**, in the sense of farming of animals in water, accounts for about 17 per cent of the total fish harvest. The bulk of fisheries is geared towards catching wild marine fish. Approximately 70 per cent of the produce from aquaculture comes from farmed finfish, the remainder being shell fish, especially mussels and oysters. Since the 1970s, there has been a dramatic increase in production of farmed fish and this trend is likely to accelerate into the 21st century. During the 10 years from 1984 to 1994, world production of farmed fish rose from just under 4 million tonnes to over 13 million tonnes. On a global scale, the bulk of aquaculture production (about 87 per cent) is concentrated in Asia (mainly China) where small-scale subsistence carp farming has proved highly successful.

Figure 6.1 A simple pond for rearing fish in a smallholding in south-west China. The pond is set in an orchard, amongst patches of vegetables and flowers.

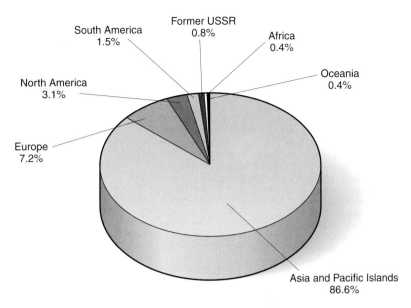

Figure 6.2 *Production of farmed fish in 1994, showing worldwide distribution. On a global scale, the bulk of the farming of finfish (about 86 per cent) is concentrated in Asia, including China, Japan, Taiwan, the Philippines and Indonesia, with less than 20 per cent in Europe and about 6 per cent in the USA. Relatively little development of fish farming has occurred in Africa and Latin America where it could make a useful contribution to food production, whereas in south east Asia, small-scale subsistence fish farming has proved highly successful*

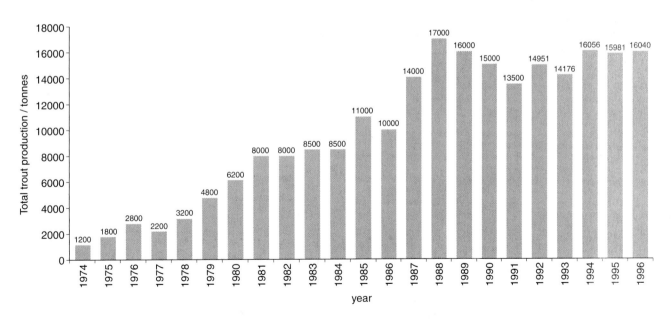

Figure 6.3 *Intensive trout farming in the UK shows a 10-fold increase since the early 1970s. By the 1990s, farmed trout production has reached a stable yet fluctuating level of around 16 000 tonnes per year. Expansion of salmon farming (nearly all in Scotland) is now competing with farmed trout and market prices for trout are falling*

FISH AND FISH FARMING

Within the fish farming industry, there is a range of systems, for marine and freshwater fish, from small-scale semi-intensive (low input) through to highly intensive. The principles involved will be illustrated by methods used for production of intensively reared freshwater rainbow trout (*Oncorhynchus mykiss*) as farmed in Britain. Fish produced in these fish farms may be marketed for human consumption in a range of forms or used to provide live fish to stock rivers and lakes for anglers.

Biology of the trout

Rainbow trout (*Oncorhynchus mykiss*) belong to the family Salmonidae, and are native to temperate north west America, though wild stocks have been introduced to favourable waters globally. In natural conditions they are carnivorous, slimly built and fast swimmers. They spawn in fresh water, but can migrate for part of their life cycle to marine waters. The rainbow trout is probably the most suitable salmonid for artificial culture, although the Atlantic salmon (*Salmo salar*) is also extensively farmed, particularly in the coastal waters of Norway, Scotland and Chile.

Fish are **poikilothermic**, so their body temperature responds directly to that of the surrounding water. This means that the rates of metabolic processes, including growth and physical activity, are strongly influenced by water temperature. It also means that energy is not used in maintaining body temperature. Because of the support provided by water, much less energy is required for support and locomotion. Fish are thus potentially more efficient than terrestrial animals at converting food into flesh.

What features of a fish, such as a trout, are linked to their life in water and how do these features differ in organisms living on land? As a start think about:
- support;
- locomotion;
- gas exchange;
- osmoregulation;
- excretion;
- temperature control.

Figure 6.4 (a) External and (b) internal features of the trout
- *the external surface has two layers – a thin outer epidermis (which is continuously sloughed off) and an inner dermis*
- *the outer epidermis contains mucus-secreting cells*
- *the mucus makes the fish more streamlined*
- *the mucus also provides some protection against entry of pollutants and microorganisms*

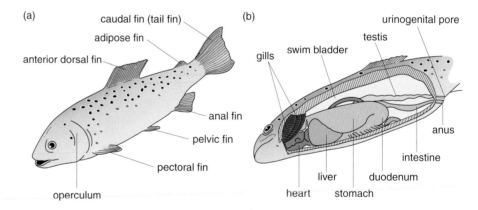

(a)
- caudal fin (tail fin)
- adipose fin
- anterior dorsal fin
- anal fin
- pelvic fin
- pectoral fin
- operculum

(b)
- urinogenital pore
- testis
- gills
- swim bladder
- anus
- intestine
- duodenum
- liver
- stomach
- heart

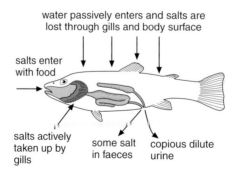

water passively enters and salts are lost through gills and body surface

salts enter with food

salts actively taken up by gills

some salt in faeces

copious dilute urine

Figure 6.5 Osmoregulation in fish – in fresh water, water is taken in by osmosis and salts are lost by diffusion; in sea water, the opposite occurs. Some exchange also occurs through the body surface. To maintain the ionic balance in fresh water, fish actively take up salts in their gills and secrete large quantities of dilute urine

Water is the medium through which **exchange of gases** occurs, as well as movement of **salts** and **excretion** from the fish. The amount of oxygen available in water is only about one-twentieth that in air and is affected by changes in temperature. Uptake of oxygen and release of carbon dioxide occurs through the **gills** which are richly supplied with blood capillaries and provide a very large surface area in contact with the water. **Ventilation** or flow of water through the gills is achieved by water entering the mouth then being forced out through the operculum due to up and down movements of the floor of the mouth. Because the gills are continuously in contact with water, movement of water and salts occurs through the gills. The main **nitrogenous waste** product is ammonia which is excreted through the gills into the water. Any deterioration of water quality, presence of particles in the water or infection from microorganisms could affect the gills, and in turn affect the respiration, maintenance of salt balance and excretion in the fish.

The body scales arise from the inner dermis. In the wild, they show growth rings of varying thickness resulting from seasonal variation, say in diet or temperature. These rings can be used to determine the age of the fish. Why do you think these differences in thickness are less evident in farmed fish?

Reproduction and breeding

In immature fish, the paired ovaries and testes (gonads) are very small and lie dorsally in the body cavity. **Maturation** and subsequent **spawning** (the release of eggs and sperm) occurs in response to specific environmental factors (such as **photoperiod** and **temperature**) that trigger a 'hormone cascade' in the fish. These environmental factors stimulate the secretion of releasing hormones from the **hypothalamus** which target the **pituitary** and cause the secretion of **gonadotrophic hormones**. These then act on the ovary and testis causing the release of further hormones (mainly oestrogens and androgens respectively). Both the ovaries and testes enlarge considerably as they mature, by which time the ovaries may occupy up to 30 per cent of the body weight and testes 10 per cent. **Ovulation** results in the release of ripe eggs into the abdominal cavity. In wild trout they would remain there until courtship behaviour led to their release through the genital pore. Similarly, **sperm** (known as **milt**) collect in the testes until released through the urinogenital opening. Fertilisation is external. During the life cycle, as the gonads mature, changes occur in the pattern of growth and nature of flesh of the fish and both sexes show a reduction in growth. There is a reduction in muscle energy reserves (fat) and pigment resulting in watery, pale flesh. This loss in quality makes it unacceptable for sale to the consumer. Males become mature at about 1 year and females at 2 years.

In the wild, spawning is an annual event, controlled by seasonal patterns of changing day length and temperature, and occurs at a time that is favourable for the subsequent development of eggs and young fish (fry). At the onset of the spawning season, trout swim upstream to shallow water where the stream bed is made up of clean stones. The female uses her tail to make a slight depression, known as a **redd**, into which she deposits her eggs. The male swims over, discharges sperm, and so the eggs become fertilised. The fertilised eggs are then covered over with gravel and left to develop.

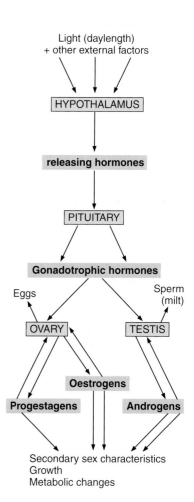

Figure 6.6 Hormones involved in the control of reproduction in fish, leading to spawning

Trout farming in operation

In intensive fish farming there is a tendency for specialisation into **hatcheries**, which produce and hatch the eggs through to young fry, and **growing-on farms**, which raise the fish (from fry) to marketable size.

Spawning and fertilisation

It is desirable to control the **reproductive cycle** so that production of marketable fish can be maintained as required, rather than being limited by the natural spawning season. Rainbow trout mature in response to short daylength, so in Britain natural production of eggs occurs from November to February. The **timing** of trout spawning can be artificially advanced or retarded by up to 4 months using photoperiod control. This involves subjecting the fish to constant long days (18 hours of light, 6 hours of darkness) and constant short days (6 hours light, 18 hours dark) at very specific periods during the maturation cycle. If artificial lighting regimes are used, they must be maintained throughout the life cycle otherwise the fish will revert to the natural spawning season. Spawning can also be induced by injection of **gonadotropic hormones**, derived from extracts of the pituitary gland. While this practice is used on a wide range of fish species, the timing can be altered by only a few weeks so its main benefit is to enable broodstock to spawn under conditions of intensive cultivation. Some farms in Britain use eggs transported by air from the southern hemisphere (for example from Tasmania or South Africa), thus taking advantage of their different spawning season.

The parent fish used to provide the eggs are known as the **broodstock**. Broodstock fish are usually kept at a lower stocking density and fed a very high quality diet to ensure good quality eggs. They are generally starved for 3 days before spawning to avoid faecal contamination of eggs or sperm. Ripe females (hens) are recognised by their swollen bellies, due to the release of eggs at ovulation into the abdominal cavity. The eggs become over-ripe if left for more than 8 days after ovulation so the hens should be sorted weekly. The eggs are removed by a process known as **stripping**. The abdomen is squeezed gently but firmly by hand, moving towards the tail. The eggs, orange in colour, flow out through the ovipositor where they are collected in a container such as a bowl or bucket. Males are stripped in a similar way to provide milt (sperm), though the timing is less critical. The milt is white and creamy in texture.

Figure 6.7 Stripping eggs from a female trout

Both eggs and sperm are kept dry at first – contamination with water at this stage will prevent fertilisation. A small quantity of sperm is then mixed with the eggs, using a finger or feather, to allow **fertilisation** to occur. The sperm enters the egg through a hole in the shell, known as the micropyle. Water is then added to the fertilised eggs and is taken in by osmosis, causing the eggs to swell or harden. This makes it more difficult for the sperm to penetrate the opening, hence the delay in adding water. The fertilised eggs are rinsed and allowed to harden. During the first half of the incubation period, the eggs are extremely sensitive to disturbance and considerable losses can occur. Unfertilised eggs and milt are less sensitive and can be transported to allow distribution to other farms. They can be stored at low temperatures (2 to 3 °C) in oxygenated conditions for a few days. Successful attempts have been made to freeze and store sperm indefinitely at −196 °C.

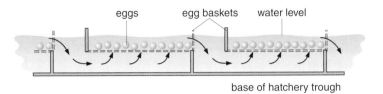

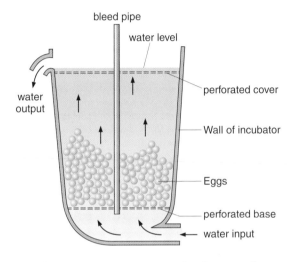

Figure 6.8 Trays containing eggs in a hatchery (top diagram and top photo in margin); the eggs rest in a perforated basket and water passes through continuously; (bottom diagram) a vertical incubator is another type of commonly used hatching system; (photo right in margin) picking out dead eggs from a hatching tray.

The fertilised eggs are placed in an incubator where they develop through to hatching. The incubator usually consists of a long, horizontal trough, holding large numbers of eggs in perforated trays. A flow of water is maintained to ensure adequate oxygenation for the developing eggs. Dead eggs, recognised by their white colour resulting from coagulation of the yolk, must be removed regularly to avoid growth of fungus. This can be done by hand or sometimes automatically. During incubation, eggs should be kept in darkness as blue or violet light is damaging to them, though orange or yellow lights are safe to use.

Development of the young trout

The time taken for eggs and young fish (fry) to develop depends on the temperature of the water and can be predicted using **degree-days**. Trout eggs require about 300 degree-days to hatch, so, at a temperature of 8 °C, eggs would hatch in about 38 days (ie 38 × 8 = 304). During incubation, eggs should continue to be kept in darkness (later semi-darkness) until the hatched fry are ready to feed. Part way through the incubation period (at about 160 degree-days for trout), the larval eyes become prominent as dark spots. At this stage the developing eggs can be safely transported from hatcheries to farms for growing on into adult fish.

Figure 6.9 Troughs for rearing young fry – a sprinkler maintains circulation of water and adequate aeration

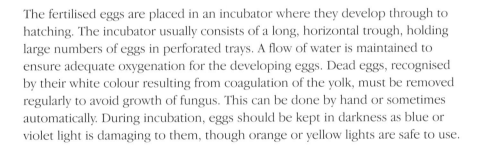

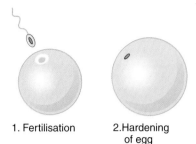

1. Fertilisation

2. Hardening of egg

3. Cell division starting

4. 'Eyed' stage

5. Hatching

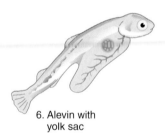

6. Alevin with yolk sac

7. Fry, able to swim and feed

Figure 6.10 Stages in development of the trout, from fertilisation of the egg to freely swimming fry (not drawn to same scale)

Newly hatched fry (also known as **alevins**) can swim through the perforations in the tray into the trough below, leaving behind the remains of the egg shells and other debris. They remain in these troughs for a few weeks, but the rate of flow of water must be increased to allow for their higher oxygen demand. At this stage, the alevins derive nourishment from the remains of the yolk sac of the egg. As this is used up they begin to swim up towards the water surface, ready to take their **first feed**. This is a critical stage for fish farmers to ensure feeding becomes established otherwise heavy losses may occur. Fry are generally reared in small tanks, less than 1 m deep, with a sloping bottom and provision for emptying and cleaning. When fry have reached a weight of approximately 5 g (200 fry per kg), they are generally sold to the growing-on farms.

Table 6.1 *Summary of events during normal egg development through to fry in rainbow trout*

Time after fertilisation (at 10 °C)	Stage	Comments
0 to 24 hours before	stripping	dry method; ripe females and males stripped into separate bowls; sperm can be pooled to give genetic mix
0	fertilisation	sperm mixed with eggs – 1 cm² sperm to 10 000 eggs
15 minutes	water hardening	excess sperm washed off; left for 40 minutes to harden
45 minutes	laying down	eggs placed in incubator
12 hours	determine fertilisation rate	sample of eggs in clearing solution (acetic acid, methanol, water); observe number that have reached 4-cell stage
7 days	determine fertilisation rate (best time)	sample into clearing solution; developing embryo seen as thick white line
16 days	eyeing	eyes seen through the shell
19 days	shocking	siphon from incubator to bucket of water; infertile eggs turn white by rupturing yolk membrane
20 days	removing infertile eggs	picked manually or mechanically; eyeing rate determined
30 days	hatching	egg shells rupture; alevins (larval fish) become free-swimming
50 days	swim up, first feeding	yolk sac absorbed; fish start feeding
approx. 130 days	fry reach 4.5 g	approximately 200 fish per kg

Growing-on farms

Ponds used for growing fish to market size are typically dug into the earth, though they may be lined with concrete, fibre glass or other materials. The size can vary considerably and an important factor is ease of access for maintenance, feeding and monitoring the fish. A convenient size is 30 × 10 m and about 1.5 m deep. If deeper, a thermocline may develop with cooler water in the lower layer which is effectively not used by the fish. A pond of this size could carry 1.5 tonnes of trout. A series of such ponds may be arranged in sequence, surrounded by walkways to allow access for feeding and cleaning, and linked to common inlet and outlet channels for water flow. Usually, the layout allows a fresh water supply to each pond. There must be some means of controlling the water level and emptying the pond at intervals for cleaning and removal of accumulated sand or silt. Screens in the form of mesh or bars are essential: at the inlet they prevent entry of wild fish, leaves and other debris and at the outlet they stop fish escaping. It may be advisable to cover ponds with netting to protect against predators.

Figure 6.11 Typical pond in a trout farm, showing provision for netting and a system for aeration of the water

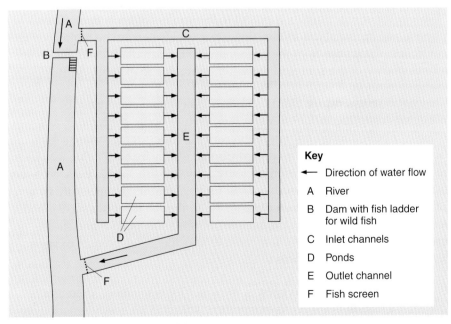

Key

→   Direction of water flow

A   River

B   Dam with fish ladder for wild fish

C   Inlet channels

D   Ponds

E   Outlet channel

F   Fish screen

Figure 6.12 Layout of ponds in a fish farm

The quality and reliability of the water source is crucial for successful growth of fish on the farm. Clear, clean water, with a neutral to alkaline pH, is preferred. The supply should continue throughout the year and ideally be gravity-fed to avoid costs of pumping. A natural spring may be suitable, provided it is tested for presence of mineral salts (which can be toxic to fish) and steps taken to ensure oxygenation as the water reaches the pond. The flow rate through the ponds needs to be adjusted to ensure adequate oxygenation, and this also influences the rate at which ammonia and carbon dioxide (from the fish metabolism) are dispersed from the water. If flow rates are too high, fish will expend unnecessary energy in maintaining their position against the current. If the flow rate is too low the water stagnates, though mechanical devices such as air blowers, and even liquid oxygen, are often used to improve the oxygen levels. The temperature of the incoming water affects both the level of oxygenation and the growth rate of the fish. If the temperature is too low, the time taken for the fish to reach suitable size extends over several years, thus increasing the cost of production. Generally it is not economic to heat the water, except during early stages of growth, unless warmed water is available, say from a factory outlet. Ultimately, it is the level of oxygenation that is probably the most critical factor in determining the stocking density, hence the profitability of the farm. The quality of effluent from the farm must be carefully monitored to ensure it is acceptable with respect to standards required by the Environment Agency or equivalent bodies.

In some situations water may be recirculated. Such systems must include ways of restoring the water to the required quality, and may involve particulate and biological or ion-exchange filtration. Despite the higher costs of such operations, they may be appropriate on a small scale or if water supplies are insufficient to meet the requirements of the fish. This expenditure may be justified where the value of the product being farmed is relatively high.

Maintenance of water quality

Maintenance of water quality is crucial to the success of a fish farm. Several physical and chemical factors affecting water quality need to be monitored and controlled: some arise from the metabolism of the fish themselves, others are a consequence of the source of the water or are due to external events.

Changes in **temperature** affect fish in different ways. Under natural conditions daily fluctuations in water temperature occur, the temperature rising to a peak during early afternoon with the lowest temperature a few hours before sunrise. Activity and growth rate of fish are both influenced by temperature. The optimum range for any particular fish species is likely to vary at different stages of the life cycle. Trout growers favour a temperature range of 5 to 20 °C, with an optimum of about 16 °C, though trout can survive under ice at 0 °C. Above 21 °C they can survive only for short periods and show extreme distress above 25 °C. Temperature also affects solubility of **oxygen** and the level of free **ammonia** in the water.

Fish depend on dissolved **oxygen** in the water for respiration. There is a diurnal fluctuation in oxygen concentration, linked to photosynthetic activity of aquatic organisms. The lowest level comes just before dawn, and oxygen concentration increases during daylight hours. For trout, a critical low level is 5 mg dissolved oxygen per dm^3 (litre). The solubility of oxygen in water decreases as the temperature rises but the demand for oxygen by a fish increases with rise in temperature. The demand also increases with higher food intake and with increased activity. As the fish grows, the demand for oxygen decreases in relation to body mass. Inadequate oxygen at any stage leads to suffocation of the fish. Adequate aeration is thus essential and the stocking density must be controlled.

Excretory products include **carbon dioxide** produced from respiration and **nitrogenous** material. The release of carbon dioxide into the water can affect pH, though normally carbon dioxide levels do not become critical. The bulk of the nitrogenous waste is ammonia and this can have serious effects on the fish. In its free state, ammonia is highly toxic, though it is far less toxic as the ammonium ion (NH_4^+). The balance between ammonium hydroxide and free ammonia depends on both pH and temperature: at a pH of 6.5 dissociation into the non-toxic ammonium ion is favoured, whereas at a pH of 8.5 there is a considerable increase in toxic free ammonia. At higher temperatures (15 °C and above) the level of toxic free ammonia increases and can very quickly become dangerous.

The preferred **pH** range of the water for trout is just below neutral to slightly alkaline (6.4 to 8.4). Fluctuations in pH cause considerable stress. Changes in pH of the water supply may result from inflow of organic acids leached from soils, though hard water (associated with limestone areas) has some buffering capacity which helps to maintain a more stable pH.

Suspended solids, when present at high levels, can cause damage to fish, particularly the gills. Such materials may arise from silt or particles in the water supply and also from fecal material in the water. If **polluted** water enters a fish farm it may be harmful to the fish. Pollutants rich in organic matter, such as sewage, farm waste or silage effluent, have a very high **biochemical oxygen demand (BOD)** which can remove oxygen from the water and cause the fish to suffocate. Algal blooms, a consequence of high nutrient levels, may also seriously deplete the available oxygen supply. This is either a result of respiration of the algae during the night or because of an increase in the BOD when the bloom dies. Death of fish from oxygen depletion is most likely during the night or early morning. Many algal species may also produce toxins or physically interfere with the fish, particularly the gills. Certain heavy metals, such as copper, lead and mercury, are toxic to fish and occur in water originating from industrial effluents. Non-metals, such as ammonia and chlorine, and complex organic compounds, including those derived from herbicides and insecticides, may also reach toxic levels. Mass mortality of fish, known as a **fish kill**, may occur quite suddenly, within a few hours of fish feeding and behaving normally. Fish kills occur most commonly from lack of oxygen, though there may be a range of underlying causes.

In pollution incidents, toxicant-specific surface changes in the gill tissue can be identified using the scanning electron microscope (Figure 6.14). This technique can be used, with chemical analysis, to confirm the identity of pollutants where negligent or accidental discharges are suspected.

Table 6.2 *Temperature and oxygen solubility*

Temperature / °C	Oxygen solubility / mg dm^{-3}
0	14.6
10	11.3
20	9.2
30	7.6
40	7.6

Figure 6.13 System for maintaining aeration in tanks in a fish farm.

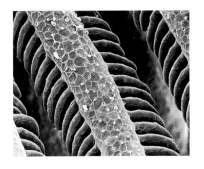

Figure 6.14 Effects of pollution on fish gills – a scanning electron microscope shows gills of a freshwater fish in (a) clean water, showing the secondary gill lamellae equally spaced; (b) polluted water with acute levels of calcium (10 mg dm^{-3}), showing characteristic 'ballooning' effects where the lamellae fill up with blood cells (magnification × 220).

FISH AND FISH FARMING

Figure 6.15 Salmon farm in inland seawater fjord in Norway

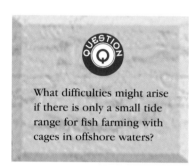

What difficulties might arise if there is only a small tide range for fish farming with cages in offshore waters?

As an alternative to ponds and tanks, fish can be grown in cages, in freshwater lakes and in the sea. Advantages of using cages for the culture of fish are that there is less likelihood of there being problems with water supply or oxygenation, though it is important that cages are strong and securely anchored in sheltered water, particularly when in the sea. There may be risk of pollution from outside sources, such as oil tanker spills which can be highly damaging, and also from the fish themselves if the water movement is insufficient or the water body is too small. Fish farm effluents must be adequately dispersed to prevent significant environmental impact. Strict controls are place on farm effluents.

Feeding and growing on to harvest

Trout are carnivores and there is a heavy demand in the diet for high quality protein. The best protein source available is fish meal, which has high digestibility and a balanced essential amino acid profile, though increasingly cheaper vegetable protein sources, such as soy, are being utilised. The ratio of protein : energy of the diet is progressively reduced as the fish grow. Typically a fry would be fed a 55 per cent protein diet while a larger fish, in excess of 50 g, would receive a 45 per cent protein diet with a higher energy level. Ideally most of the dietary protein is used for growth with lipid supplying the energy requirement. The ability of trout to utilise dietary carbohydrate is very poor and so only low levels are included, usually under 15 per cent. A vitamin and mineral premix is also supplied in the diet.

Feed rates depend on the size of the fish and the temperature of the water. In relation to body weight, relatively higher rates of feeding are required by first-feed fry, and the rate decreases as the fish grow larger. Feeds are usually supplied as pellets, the size being adjusted as the fish grow. The feed can be delivered by hand or automatically. Young fry require almost continual feed, in fine pellets, whereas mature broodstock fish can take their whole day's ration at once, as large pellets. The cost of feed is a relatively high proportion of production costs, so it is important to avoid wastage.

The red or pink colour of the flesh in salmonids is due to fat-soluble carotenoid pigments. In the wild, these are obtained from eating other animals, such as crustaceans. In response to consumer demand for pink-coloured flesh in both farmed trout and salmon, this colour is achieved by the inclusion in the diet of the pigments canthaxanthin and astaxanthin. These are artificially synthesised but are identical to the naturally occurring pigments. Of these two pigments, astaxanthin is preferred as it is predominant in wild salmonids. The cost of flesh pigmentation is high, typically representing 10 to 20 per cent of the diet cost.

Within a pond, fish feed and grow at different rates so it may be necessary to grade and sort fish at intervals during their growth. This allows fish of about the same size to be reared together. At harvest it is important to provide uniform sizes to suit the retail market. Frequent grading, particularly of the fry, is avoided as disturbance leads to a reduced growth rate. Fish are collected by a sweep net into a small area of the pond then passed through bars set at a

Figure 6.16 Trout passing through a grader in a fish farm.

distance to allow certain body sizes through. Automatic graders are sometimes used, particularly when selecting fish for harvest.

In most of Europe, trout are usually prepared for marketing at between 180 and 280 g, though some much larger fish (2 kg or more) are sold, particularly when used for smoking. Typically in the UK, the 180 to 280 g size is achieved at the age of about 10 to 14 months, depending on temperature. Care must be taken in handling the fish at capture to avoid damage and bruising because such fish are then unsightly and down grading occurs. This is important with fish reared both for angling and for food. Before harvest, fish are starved for up to 48 hours as this empties the gut and prolongs the storage properties of the fish. The gutting, cleaning and transport of harvested fish is now linked to mass markets and must be carried out quickly and efficiently to retain the fresh quality of the fish flesh.

Figure 6.17 Farmed trout and salmon in the supermarket

Control of sexes in fish stocks

The flesh of mature fish tends to lose quality and after spawning there is a high mortality rate. Males mature before females and tend to become aggressive at maturity. Fish raised for food are generally harvested before reaching sexual maturity to avoid these problems. However, it is relatively easy to manipulate the sex of fish to produce single-sex or sterile stocks and these techniques are now actively used in commercial fish farms. In trout farming, stocks consisting wholly of females are preferred.

In trout, females are homogametic (**XX**) and males heterogametic (**XY**). The genetic sex is established at fertilisation but determination of the actual sex occurs later in the development in response to hormone levels during the period of sexual differentiation. Sex reversal can be achieved by feeding hormones, thus overriding the genetic sex (as predicted by the **X** and **Y** chromosomes). Male hormones or androgens will produce all males (with testes and male sexual characteristics at maturity) and oestrogens will produce fish with ovaries and female characteristics. The hormones are administered in the feed to first-feeding fry during the period of sexual differentiation. This technique is rarely used for fish intended for human food because of consumer objections to hormone-treated fish.

Sex reversal can be extended to the second generation of fish to produce all female fish. Genetic females (**XX**) are **masculinised** by treatment with male hormones. All sperm then contains an **X** chromosome, rather than having a mixture of **X** and **Y**. Fertilisation of **X** eggs with **X** sperm then results in all female (**XX**) offspring. Although there are some practical difficulties in obtaining sperm from the masculinised females, this technique accounts for over 90 per cent of farmed trout and avoids feeding hormones to production fish.

Losses associated with the onset of sexual maturity can be avoided by **sterilisation**, which allows the fish to continue their growth beyond the size at which they would normally mature. Sterilisation can be achieved by hormone treatment, supplying higher doses of male hormones in the feed. Alternatively, changes in the chromosome number can be induced leading to

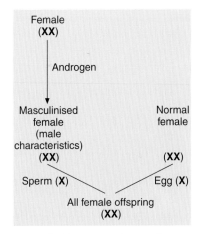

Figure 6.18 Salmon fillets being smoked in Scotland.

Female
(**XX**)

↓ Androgen

Masculinised female (male characteristics) (**XX**)   Normal female (**XX**)

Sperm (**X**)   Egg (**X**)

All female offspring (**XX**)

Figure 6.19 Sex reversal to produce stocks of all female fish

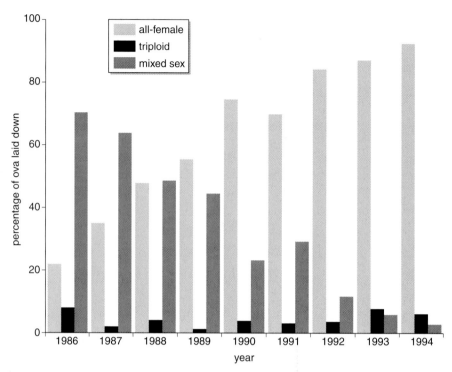

Figure 6.20 Manipulation of the sexes to produce females and use of triploids in intensive fish farming has shown a marked increase over the past decade (data from Scotland)

production of triploids. This can be done by subjecting the eggs to an environmental shock – either pressure or temperature – at a critical stage just after fertilisation. In trout, this is achieved by applying the shock 30 to 40 minutes after fertilisation at 10 °C. The final meiotic division of the egg nucleus fails to occur, resulting in a diploid (2n) egg nucleus which becomes triploid (3n) when fertilised by the haploid sperm. Up until the maturity of normal diploids, triploid trout show no growth improvement, However, from the age of 2 years, because they do not mature, they show improved growth and avoid the carcass quality problems associated with maturation.

The honey bee

Honey bees have not been domesticated in the same way as cattle, sheep and chickens, but there has been a long-standing association with human communities and exploitation of their activities in the production of human food. In the Bronze Age, honey and other products were collected from wild bees and in medieval times, beekeepers tended colonies of wild bees. The position of beekeeper was of importance, the holder having the privilege of carrying a crossbow in case of attack by bears while collecting the honey. The practice of moving hollow trees, containing colonies of bees, nearer to dwellings gave rise to the practice of **apiculture**, or bee-keeping, making it safer and easier to collect the honey.

Honey is an easily digestible food and was the only sweetener available in any quantity before the extraction of sugar from sugar cane and sugar beet was developed. Since honey ferments very readily, the remains of a honeycomb soaked in water produced an intoxicating beverage, which became known as mead. This drink was popular for thousands of years, particularly in countries such as Britain, where grapes could not be grown easily. Until the 16th century, much of the honey production in Britain took place in monasteries, where bees were kept primarily for their wax, which was made into candles.

Reproduction and life cycle

There are some 20 000 known species of bees, most of which are solitary, but members of the genus *Apis*, the honey bees, are social insects, living in permanent colonies where there is a high degree of organisation and caste differentiation. Each colony consists of a single, reproductive female, the **queen**, a large number of sterile females, known as **workers**, and a much smaller number of males, called **drones**. In the wild, *Apis mellifera* tends to build nests in enclosed sites and each nest is composed of a series of parallel combs. This species can adapt to using hives and will build bigger combs storing larger amounts of honey. *A. mellifera* is found throughout Africa, the Middle East and temperate regions of Europe and it has been introduced into North and South America and Australasia. The large stores of honey and pollen that are built up, together with the preference of this species for building nests in enclosed spaces, enables survival during cold winters in temperate regions.

A typical colony of honey bees in the summer may consist of a single queen, a few hundred drones and around 50 000 workers. Within the hive, there are stores of honey and pollen, together with a number of combs containing bees at all stages of their development. The part of the hive in which the young bees develop is called the brood nest and it is here that the queen is found laying eggs into the cells. The **brood nest** is usually surrounded by cells containing pollen, with the cells containing honey around and above the pollen-storing cells.

THE HONEY BEE

The three different castes of honey bees have distinct differences in size and structure, making it relatively easy to distinguish them. These differences are shown in Figure 7.1 and summarised in Table 7.1.

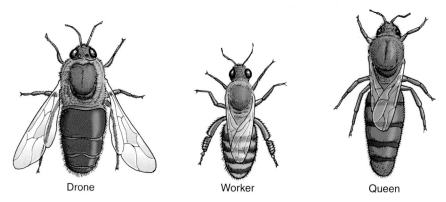

Drone   Worker   Queen

Figure 7.1 Honey bee castes

Table 7.1 *Differences between different castes of honey bee*

Drones	Workers	Queens
fertile males	sterile females	fertile workers
big, broad body	smaller body than drone or queen	larger, longer body than workers
well-developed wings	small wings	shorter wings than workers
no sting	barbed sting	unbarbed sting
reduced mouthparts	mouthparts modified for sucking up nectar and moulding wax	poorly developed mouthparts; fed by worker bees
only function is to mate with queen	variety of functions within hive and as forager	function is to lay eggs and swarming
lives 4–5 weeks; killed or driven out of hive in autumn	lives 4–5 weeks in summer, longer if over winter	may live 5–6 years

Drones exist only to mate. They do not perform any other useful functions and most will die before mating because they get too old or the workers push them out of the hive. Those drones that do succeed in mating with the queen die immediately after copulation, because their abdomen and genitalia are ruptured by the process of mating. Drone production can be controlled and they are only produced within a colony when there are likely to be queens available for mating. In the spring, drones are usually reared just before the emergence of virgin queens, production peaking about 4 weeks before swarming. Fewer drones are produced during the summer, when fewer queens are produced, but there is often a slight rise in late summer, which coincides with the August peak in swarming.

Once the drones emerge from their pupal stage, they are initially fed by the workers and then begin to feed themselves from the honey cells. Their reproductive organs continue to develop for about 12 days after emergence, during which time they rest in the hive and feed. They begin to leave the hive on orientation flights, lasting a few minutes. Mating flights last longer, up to an hour, depending on the weather conditions.

Queens only mate during the early period of their lives, before beginning to lay eggs. Virgin queens become sexually mature about 5 or 6 days after their emergence from the pupal stage. They too undergo one or two short orientation flights before the mating flights. Such flights usually take place in mid-afternoon on sunny days over a period of 2 to 4 days. The most favourable conditions for mating flights are winds of less than 20 km per hour, little cloud cover and a temperature higher than 20 °C.

Mating takes place at special sites, called congregation areas, where drones collect and fly around awaiting the arrival of virgin queens. Drones and queens from several different nests or hives may gather in such areas, so there is likely to be considerable mixing of populations. It is thought that the drones release a chemical substance, called a pheromone, which attracts other drones to a congregation area and then, later, the virgin queens.

Copulation is rapid and once contact has been made between a drone and a queen, the actual mating process may be completed in less than a couple of seconds. The queen opens her sting chamber, the drone inserts his endophallus and the sperm is ejaculated into the queen's oviducts as the drone flips backwards. The sperm are stored in the queen's spermatheca, which can store up to 7 million sperm. As mentioned earlier, the drone dies soon after copulation, but the queen usually mates with more than one drone on a mating flight. One queen was found to have mated with 17 different drones on a single flight. It can take up to 4 years for all the sperm stored in a queen's spermatheca to be used up.

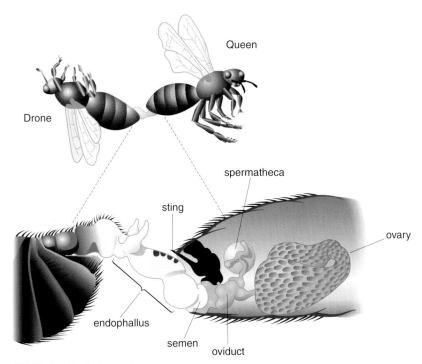

Figure 7.2 Mating in the honey bee

The cylindrical eggs, between 1.3 and 1.8 mm long and pearly-white in colour, are slightly curved with the head end thicker than the abdominal end. As the queen lays the eggs, one per cell, she glues them to the floor of the cell at one end. The queen can lay eggs at the rate of 2000 a day. She can lay two types of eggs, fertilised and unfertilised. The unfertilised eggs develop into drones, which are haploid male bees, capable of producing haploid male gametes by mitosis when they are mature. The fertilised eggs develop into diploid female bees. Whether the fertilised eggs develop into fertile queens or sterile workers depends on the type of cell into which the egg is laid and on how the larva is fed. An egg laid into a worker cell can be moved to a queen cell where it will develop into a queen, given the correct type of food. The reverse can happen: an egg laid into a queen cell can be moved to a worker cell and develop into a worker.

All the larvae, whether they develop from fertilised or unfertilised eggs, are fed on **brood food**, which is produced from the hypopharyngeal and mandibular glands of the nurse bees. Pollen and nectar may be added to the diet of the worker larvae and the drone larvae after the third day. The secretions from the mandibular glands of the nurse bees are white, whereas those from the hypopharyngeal glands are clear. Yellow secretions are derived from pollen. The brood food fed to the worker larvae has a ratio of 2 white : 9 clear : 3 yellow secretions; that fed to the queen larvae has a ratio of 1 white : 1 clear, with no yellow secretion. In addition, the quantity of food given to the queen larvae is greater, so they grow bigger. In the same way that eggs can be switched between queen cells and worker cells, so can larvae. Worker larvae fed on brood food high in mandibular secretions develop into queens.

The rate of development of the young bees depends on temperature and nutrition, but it takes from 14 to 17 days for a queen to develop from an egg, 16 to 24 days for a worker and 20 to 28 days for a drone. If the temperature falls below 35 °C in the hive at any stage, then the time for development will be prolonged. Underfeeding of the larvae can also result in longer development times. The timing of each stage of development for the different castes is given in Table 7.2.

Table 7.2 *Stages of development of the different castes of honey bees*

Stage of development	Queen	Worker	Drone
incubation of eggs	3 days	3 days	3 days
duration of larval development in uncapped cells: larvae fed by workers	3 to 5 days	5 to 6 days	4 to 7 days
duration of pre-pupal stage: cells capped with wax; cocoon formation	3 to 4 days	3 to 5 days	4 to 6 days
duration of pupal stage	4 to 5 days	8 to 9 days	8 to 9 days

At the end of the pupal stage, there is a final moult before the adult bee emerges. The emerged bee has a soft cuticle, which hardens during the next 12 to 24 hours. During the next few days, internal development is completed. The worker bees have a complex timetable of activities which varies according to their age. The schedule is not a rigid one as older bees have been seen carrying out tasks usually done by the younger workers. In order to obtain the nutrition necessary to carry out the various tasks, the young workers must begin to feed and begin consuming pollen within a few hours of emerging from the pupa. A timetable of the activities of the worker bees is given in Table 7.3.

Table 7.3 *Timetable of activities of worker bees*

Days after hatching	Activities in the hive	Activities outside the hive
1 to 3	fed by other workers; cleans out cells of recently-hatched bees	
3 to 5 5 to 12	feeds older larvae on pollen and honey hypopharyngeal glands secrete brood food (royal jelly); feeds young larvae and queens; helps to keep brood warm	
12 to 20	wax glands on abdomen become active; can secrete wax for comb building and repair; collects nectar and pollen from foraging bees; processes nectar. cleans hive; may act as a guard bee	begins to leave hive for short flights
21 to 40	communicates position of sources of nectar to other foraging bees using 'round' and 'waggle' dances on the combs	becomes a foraging bee; daylight hours spent collecting water, nectar, pollen and propolis

Workers live for about 6 weeks, but many of those that emerge in the autumn will probably survive until the following spring.

The queen appears to control the activities of the workers by producing chemical substances, called **pheromones**. Some of these substances are secreted from her mandibular glands and make up what is usually referred to as **queen substance**. Virgin queens produce small quantities when they first emerge, but secretion increases with age and may vary throughout the year. Older queens do secrete less, particularly when their laying ability begins to fail. Queen substance has been shown to:
• prevent the rearing of queens by the workers, which will in turn prevent reproduction by swarming
• inhibit ovary development in workers
• attract drones for mating.

When swarming does occur, the queen's pheromones attract workers to the cluster, stabilise the cluster and help in the movement to a new nest. The swarm will only move as a group if the queen is present.

It is clear that the pheromones produced by the queen have a profound influence on the behaviour of the workers and on the functioning of the whole colony, so it is of interest to consider how the pheromones are spread through the colony. The queen is groomed by workers, who pick up the substances on their bodies and their antennae. The workers then groom themselves, transferring the substances to other parts of their bodies and also ingesting some. Workers that have been in contact with the queen make contact with other workers, transferring pheromones via antennal contact, thus distributing the substances throughout the colony.

Other pheromones are produced by the workers and the drones. Some enable the members of one colony to recognise each other. Beekeepers know that if they introduce a 'foreign' queen into a hive, she must be protected for a few days, otherwise she is killed very quickly by the workers.

Swarming is a form of reproduction in the life of a honey bee colony. It involves division of the colony, whereby the majority of the workers, together with a queen, leave the nest and search for a new location. The queen may be old or may be a newly emerged virgin queen. The advantage of swarming is that the queens get assistance from the workers in building a new nest, rearing new workers and foraging for food. The trigger for swarming appears to be overcrowding or congestion of the hive, a large number of young workers and the reduced transmission of queen substance, a combination of factors which induce queen-rearing.

Adult populations of honey bees gradually decline during the winter, with minimum numbers in March. Brood-rearing begins during the late winter and early spring and about 2 to 4 weeks before swarming, new queen cells appear. Swarming is most likely during May and June, although some swarms have been recorded in April. Swarming usually occurs as the queen larvae begin to pupate, so that at least one virgin queen is ready to emerge afterwards. After the first swarm has left, more workers are reared and numbers build up again.

Bees are kept commercially in **hives**, where the environment can be controlled so that the maximum amount of honey can be produced for human consumption. There are many different designs around the world, but the two most commonly seen in Britain are the National and the WBC. The WBC was designed by William Broughton Carr and can be distinguished from the National by its characteristic outer walls, or 'lifts', often painted white. Both hives have similar characteristics such as:
* blocks to raise it off the ground and prevent damp
* a floor containing an entrance block
* a brood box with frames, providing the main home of the colony
* a queen excluder, which prevents the queen from laying eggs in the food stores above
* one or more supers, in which the food, pollen and honey, is stored
* a roof to protect the hive from the weather.

The supers fit on top of the brood box and contain the frames which act as the food and honey store. Frames for the brood box consist of four-sided wooden supports with a wax foundation on which the workers construct the cells. The workers build out the cells until they are the correct size for the queen to lay the eggs in. Similar frames are used in the supers and the workers construct cells in which food and honey can be stored. In the brood box and in the supers, the frames are held in place, at the correct distance apart, by removable metal or plastic ends.

Both types of hive illustrated can be moved from one site to another relatively easily without the combs breaking.

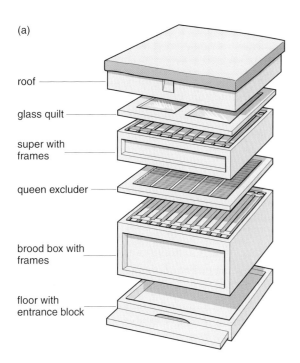

(a)

- roof
- glass quilt
- super with frames
- queen excluder
- brood box with frames
- floor with entrance block

(b)

Figure 7.3 (a) a National hive and (b) a WBC hive, showing the component parts

Economic significance of the honey bee

In North America, China, Australasia and parts of Europe, honey production is a highly developed industry and the rearing of large colonies of bees is big business. In developing countries, apiculture is a useful form of agriculture. It is not expensive to set up and can provide valuable food and a source of income. A wild swarm can be collected, hives may be constructed from local materials and then located on land which is unsuitable for other purposes.

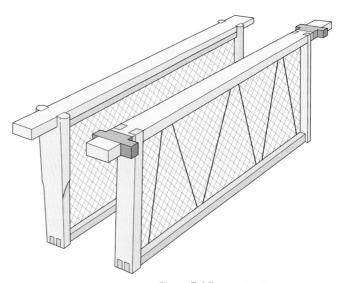

Figure 7.4 Frame structure

Honey is made mainly from **nectar** collected by the foraging worker bees. Nectar is a solution of sucrose, secreted by glandular tissue in some insect-pollinated flowers. The areas of glandular tissue are called **nectaries** and are usually situated at the base of the petals or on the receptacles of the flowers. Visiting bees, often guided by markings or textures on the petals of the flowers, push past the male and female reproductive structures in order to get to the nectar and in doing so, transfer pollen from a previously visited flower on to the stigma. The nectar is sucked up by the mouthparts of the bee, which then withdraws, brushing past the reproductive structures once again. Pollen from the anthers of the flower sticks to the hairs on the bee's body. Foraging bees may visit a large number of flowers of the same species, transferring pollen from one to another and bringing about **cross-pollination**. Cross-pollination is important for some crops, particularly in the commercial production of apples, and it is not unusual for hives to be transported into orchards during the flowering period to ensure a good crop. Many other fruit crops, such as blackberries, cherries, strawberries and pears, depend on bees as pollinators. Pollination by bees is also important in crops such as cabbage, broccoli and carrots, which are not grown for their fruits, but where it is necessary to maintain supplies of seeds. Bees have been used in glasshouses to pollinate cucumbers.

The nectar is taken into the honey sac, where it is stored until the bee returns to the hive. It may be regurgitated into a cell on the comb or passed to another bee. The bees repeatedly swallow the nectar, mix it with enzymes and regurgitate it, until the enzyme action and the evaporation of water results in its conversion into **honey**. The enzymes, present in the saliva of the bee, are secreted from the hypopharyngeal glands and include sucrase, which catalyses the hydrolysis of sucrose to glucose and fructose, and glucose oxidase, which oxidises glucose to gluconic acid. Gluconic acid helps to keep the pH of the honey low, so that the growth of bacteria is inhibited. The honey is deposited into cells and further evaporation is achieved by fanning, so that the water content is reduced to less than 18 per cent. When enzyme activity ceases and the water content is low, the cells are capped with wax until the honey is required for feeding.

Foraging bees also collect pollen for its value as a source of protein for the colony. As has already been described, the bee's body becomes covered with pollen as it moves past the anthers of the flowers visited. This pollen is brushed off, using the hairs on the legs, and packed into special pouches, called pollen baskets, on the third pair of legs. On returning to the hive, it is packed into cells together with an acid, which prevents germination and bacterial activity. It is possible that the workers add enzymes to the pollen as it is being packed into the cells and some preliminary digestion may take place. When the pollen has been processed for storage, it is called 'bee bread' and is ready for ingestion and digestion by the bees.

It has been known for a long time that worker bees can communicate with one another, especially about the location of rich food sources. Some investigations carried out by Karl von Frisch, and described in his book *The Dance Language and Orientation of Bees* in 1967, indicated that there were two distinct dances performed by the foraging bees when they returned to the hive. He suggested that these dances gave other bees information about the direction and distance from the hive of the food source that they had visited. The simplest dance, the round dance, appears to inform other workers that there is a source of food close to the hive, less than 15 m away. The waggle dance is used to communicate information about food sources more than 100 m away. The characteristics of the dance provide information about the direction of the food source as well as its distance from the hive and its quality. It appears that there are transitions between the round dance and the waggle dance that are used for food sources between 15 m and 100 m from the hive. These dances are illustrated in Figure 7.6.

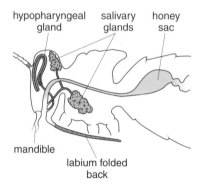

Figure 7.5 Section through part of a worker bee to show glands and honey sac

(a) round dance   indicates that the source of food is within a radius of 80 m

worker bee moves rapidly round in a circle, first to the left, then to the right

The faster the dance, the richer the source of food and the closer it is to the hive

(b) waggle dance

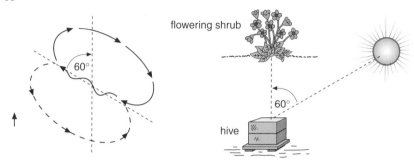

flowering shrub

60°

60°

hive

This dance indicates food is further away. The tail end is 'waggled' on the straight part of the dance; the slower the wagging, the further away the food source.
The location of the food source is indicated by the angle of the straight part of the dance on the vertical surface of the comb, which corresponds to the angle between the sun, the hive and the food source.

Figure 7.6 Round dance and waggle dance

As mentioned earlier in this chapter, bees were kept by monks, mainly for the production of wax for candles. The wax-producing glands on the underside of the abdomen of the worker bees reach their maximum size and activity between 5 and 15 days after emergence. The wax, largely composed of hydrocarbons, is produced by metabolising honey in special fat cells associated with the glands. It is secreted as a liquid and hardens into flakes, which are removed by the hind legs of the worker and passed forwards to the front legs and the mandibles. The wax is mixed with saliva and then kneaded into the right consistency, so that it can be moulded into shape. During comb-building, the temperature within the hive is kept at about 35 °C and the workers keep together in tight clusters.

The beeswax extracted from hives is usually white, although it may have a slight yellowish tinge due to carotenoid pigments in pollen. It is used for candles, cosmetics, polishes and waterproofing of textiles. Other marketable products from hives include bee venom, royal jelly and propolis. Royal jelly, which is secreted from the mandibular and hypopharyngeal glands of the young workers, is sold as a health food. It contains vitamin B_5 and has been used in the cosmetics industry in a number of different preparations. Propolis, derived from plant resins collected by the workers, is a sticky brown substance, which is used to seal cracks in the hive.

PRACTICALS

Effects of different nutrient levels on the growth of a crop plant

Introduction

The aim of this practical is to investigate the effects of different nutrient concentrations on the growth of a crop plant. Solutions of potassium nitrate are used to supply the plants with nutrients and the effect on plant growth and yield assessed after a suitable time. One important principle of experimental design in this practical is the use of a Latin square arrangement for the replicates. A Latin square is a method of randomising the positions of the replicates in order to avoid so-called edge effects. Suppose, for example, that all the high nitrogen plots were arranged along one side of a square. In this case, all the high nitrogen plots might receive more light then other treatments, which will influence the results. You will see from the Latin square arrangement below (Figure P.1) that each row and each column contains one of each treatment. This arrangement gives three replicates, but Latin squares can be used with four or more replicates.

Columns

N1	N2	N3
N3	N1	N2
N2	N3	N1

Rows

Figure P.1 A Latin square arrangement for three replicates of three treatments, N1, N2 and N3. The three treatments indicate different levels of a nitrate fertiliser

Materials

- Plastic seed trays or other suitable containers
- Levington's compost, or similar
- Supply of wheat or barley grains
- Potassium nitrate solutions:
    - N1 = 0.25 g KNO_3 per dm^3
    - N2 = 1.0 g KNO_3 per dm^3
    - N3 = 2.0 g KNO_3 per dm^3
    - *NB: Potassium nitrate (V) is an oxidising agent. Contact with combustible material may cause fire*
- Balance to record the yield

OXIDISIN
potassiur
nitrate(V

Method

1. Fill the seed trays with compost, and label each one suitably.
2. Sow the cereal grains in each tray at a density of one per square centimetre. It is advisable to sow two grains at each position, then thin out, if necessary, to give one plant per square centimetre. Seeds may have been treated with a fungicide. Ensure hands are thoroughly washed after handling seeds, or wear disposable gloves.
3. Arrange the trays in a Latin square, as shown in Figure P.1.
4. Water the trays regularly with the appropriate solutions of potassium nitrate. Ensure that each tray receives the same volume of solution.
5. When the plants have grown sufficiently, harvest each tray by cutting the plants off at ground level and record the yield as fresh mass in each.

Results and discussion

1. Record your results in a table, showing the total yield for each of the three treatments.
2. Plot a bar graph to show the total yield for each of the three nitrate levels.
3. Comment fully on your results. Are the differences in yield significant?

Suggestions for further work

1. You could develop this practical to investigate other factors affecting the yield of crop plants, such as density (numbers of plants per cm^2) both with and without nitrate fertiliser.
2. Investigate the relationship between plant density and the number of tillers formed.

3 If you are able to obtain samples of different varieties of barley, compare the effects of nitrate fertilisers on the yield of two different varieties.

Effects of competition from weeds on the yield of a crop plant

Introduction

Radish is very similar to some of the weeds which infest fields of barley and compete with it for light, space, water or nutrients. In this investigation, radish is used to investigate the effect of competition on the growth and yield of barley. Pots containing barley and radish only are used as controls with which to compare the yield of barley when competing with radish.

Materials

- Nine 12 cm diameter flower pots
- Barley grains and radish seeds
- Sowing guide, illustrated in Figure P.2
- Suitable compost, such as Levington's sowing compost
- Top pan balance

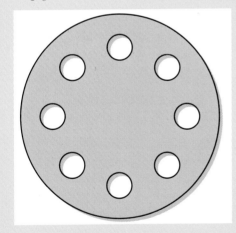

Figure P.2 A sowing guide, cut from a piece of card. This is used to ensure that seeds are sown with the same spacing in each pot

Method

1 Fill the flower pots with compost, then label suitably.
2 Use a sowing guide to sow 16 barley grains (two per position) in each of three pots of compost.

3 Repeat this procedure with radish seeds in three more pots of compost.
4 Sow three further pots with a mixture of barley and radish, arranged alternately. Again, sow two seeds per position. *Wash hands after handling seeds, or wear disposable gloves.*
5 Position the pots in a Latin square arrangement (see Practical: *Effects of different nutrient levels on the growth of a crop plant*).
6 Keep the pots watered; thin the seedlings to one per position.
7 Leave to grow for a suitable time, then harvest the plants in each pot. Record the fresh mass of barley and radish plants separately. Record the number of tillers produced on each of the barley plants.

Results and discussion

1 Record all your results in a suitable table, to show the mean fresh mass of barley and radish plants when (i) grown separately, and (ii) grown together.
2 Compare the yields of barley grown separately and in competition with radish, and suggest reasons for any differences there may be.
3 What effect did competition have on the number of tillers produced by barley?
4 Consider the implications for farmers of the results of this investigation.

Suggestions for further work

1 Investigate the effect of growing the plants in different sized pots, which will influence root competition. For example, you could try pots with diameters of 8 cm, 12 cm and 20 cm.
2 Find out which species of weeds grow in cereal crops. If seeds are available, you could investigate the effect of competition between other species of plants.

PRACTICALS

Propagation of plants by cuttings

Introduction

A cutting is part of a plant, such as a shoot or a leaf, which is removed and induced to form roots and to become an individual plant. Cuttings provide a means by which plants can be propagated, for example, to increase numbers of plants with desirable characteristics. **Stem cuttings** provide the usual means for propagating many species of shrubs and glasshouse plants. The formation of adventitious roots, that is, roots which arise directly from the stem, may be increased by treatment of the cut end of the stem with **auxin**. Some plants, including *Coleus* and willows, will root readily without applied auxins, but others will not do so unless treated with auxin. A few species will not root even when treated with auxin.

Materials

- Flower pots
- Polythene bags and elastic bands
- Suitable compost
- Stem cuttings from suitable plants, such as *Coleus*, *Erica*, *Fuchsia*, mint or willow.
- Rooting powder (commercial rooting powders contain synthetic auxins, usually indole butyric acid, IBA, or α naphthalene acetic acid, NAA). *Safety: Follow all precautions written on the rooting powder container. If skin is accidentally contaminated, wash off the preparation immediately. Wash hands after completing the procedure.*

Method

1 Fill flower pots with a suitable compost. Trim the cuttings by removing their lower leaves and cutting the stem cleanly just below a node.
2 Dip the end of the cutting in rooting powder, tap off the excess, then inset the cutting into the compost. Several cuttings can be placed around the edge of one pot.
3 Cover with a polythene bag and secure with an elastic band.
4 Keep the cuttings moist and shaded from direct sunlight.

5 When rooted, the cuttings should be transplanted and kept shaded for a few days until they regain full turgor.

Suggestions for further work

1 Investigate the effectiveness of rooting powder in stimulating adventitious root development, using treated and untreated cuttings of the same species.
2 Investigate the relationship between the number of roots formed per cutting and the number of leaves on the cutting.
3 Investigate the effect of auxin concentration on the development of adventitious roots. Use young plants of the French bean (*Phaseolus vulgaris*), cut off 1 cm above soil level. Mung bean (*Vigna radiata*) seedlings are also suitable. Mung beans are germinated in vermiculite and cuttings taken which consist of the terminal bud, two primary leaves, epicotyl and 4 cm of the hypocotyl. The cotyledons are carefully removed before the cuttings are placed into the treatment solutions.

Place separately in conical flasks containing
  (i)   distilled water
  (ii) auxin solution, 5 mg per dm^3
  (iii) auxin solution, 50 mg per dm^3.
Leave on a window sill for at least one week, then examine. Note any development of thickening of the stem, or adventitious roots.

An introduction to the propagation of plants by tissue culture

Introduction

Plant tissue culture involves the growth of isolated cells or tissues in controlled, aseptic conditions. It is possible to use plant tissue culture to regenerate whole plants, a technique referred to as **micropropagation**. One of the uses of this technique is to propagate rare or endangered species which are difficult to propagate using conventional methods of plant breeding. Micropropagation is also used to produce ornamental plants on a large scale for commercial purposes, including pot plants, cut flowers and orchids. The techniques of plant tissue culture are also used to eliminate pathogens from infected plants, for example in the production of virus-free plants, such as carnations, cauliflowers and potatoes. There are a number of different types of plant tissue culture, including:

- **embryo culture**, cultures of isolated plant embryos
- **organ culture**, cultures of isolated organs including root tips, stem tips, leaf buds and immature fruits
- **callus cultures**, which arise from the disorganised growth of cells derived from segments of plant organs, such as roots.

The isolated part of the plant used for culture is referred to as the **explant**, which can be almost any part of the plant. The tissue used as the explant is grown in or on culture media, containing a variety of mineral nutrients, plant growth regulators, such as auxins and cytokinins, sucrose, and amino acids.

In this practical, apical meristematic tissue is removed from a cauliflower and transferred to a growth medium. Under appropriate culture conditions, these meristems will change from a floral to a vegetative pattern of growth and can be induced to grow into intact plants. Careful aseptic technique is required to surface sterilise the tissue and to avoid subsequent contamination when setting up the cultures. Details of making an aseptic transfer are given in *Microorganisms and Biotechnology*.

Materials

IRRITANT
sodium
hypochlorite
solution

CORROSIVE
concentrated
sodium
hypochlorite

- Containers of cauliflower meristem culture medium (obtainable from Philip Harris Ltd)
- Fresh cauliflower
- Screw-capped bottles containing 10 per cent sodium hypochlorite solution
- Screw-capped bottles containing about 10 cm³ of sterile distilled water
- Sterile scalpel
- Sterile forceps or bacteriological loop

Method

1 Remove a cauliflower floret and use a sterile scalpel to score the surface into squares about 4 mm × 4 mm.
2 Carefully cut off several pieces from the outer surface of the cauliflower, to a depth of about 4 mm, and transfer these into a container of 10 per cent sodium hypochlorite solution.
3 Cover the container and shake vigorously for 1 minute. Leave to stand for 10 minutes to surface sterilise the tissue.
4 Transfer the explants to a container of sterile distilled water, using sterile forceps or a sterile bacteriological loop.
5 Shake the explants carefully for 1 minute.
6 Transfer the explants to a second container of sterile distilled water, again shake for 1 minute.
7 Repeat step 6 with a third container of sterile distilled water.
8 Finally, transfer one explant aseptically to each container of growth medium. Label and date your cultures; leave in the light at about 25 °C.

Results and discussion

1 Report on the growth of your meristem cultures.
2 Find out the composition of a plant tissue culture medium such as White's medium, or Murashige and Skoog (MS) medium. How are tissue culture media sterilised?

Suggestions for further work

Further suitable practicals involving plant tissue culture can be found in:
Biological Sciences Review, Volume 10, Number 3, January 1998, published by Philip Allan;
Practical Biotechnology – a guide for schools and colleges, 1993, published by the National Centre for Biotechnology Education, University of Reading;
Fuller, M.P. and Fuller F.M. (1995) Plant tissue culture using Brassica seedlings *Journal of Biological Education*, **29**(1), 53–59.

Determination of the viability of pollen grains

Introduction

Pollen grains can be germinated *in vitro* by placing them in a suitable solution. Pollen grains of some species may germinate in water, but the percentage germination is low, the pollen tubes grow slowly and may burst. The percentage germination and growth of the pollen tube can be increased by using sucrose solution. Sucrose acts as a nutrient and reduces osmotic effects, preventing the tubes from bursting. Boron appears to be essential for successful pollen tube growth in some species, so boric acid is added to the culture medium. The aim of this experiment is to investigate pollen viability by culturing pollen grains in sucrose solutions.

Materials

- Flowers, such as *Lilium*, *Narcissus*, or tulips
- Microscope slides and coverslips
- Sucrose solutions: 0.2, 0.3, 0.4 and 0.5 mol per dm³ made up in distilled water containing 0.01 g per dm³ boric acid (trioxoboric (III) acid)
- Plasticine
- Microscope

Method

1 Place a ring of plasticine on a microscope slide, so that the coverslip will be supported about 4 mm above the surface of the slide.
2 Place one drop of a sucrose solution in the centre of a coverslip, and dust a small amount of pollen onto the sucrose solution.
3 Invert the coverslip onto the ring of plasticine, so forming a **hanging drop preparation**.
4 Leave the preparation at a temperature of about 20 to 25 °C and examine using a microscope after 1 to 2 hours.
5 Count the total number of pollen grains visible in the field of view and the number which have germinated.

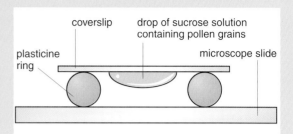

Figure P.3 A hanging drop preparation to investigate the germination of pollen grains

Results and discussion

1 Record your results in a table, showing the percentage of pollen grains which have germinated, that is percentage viability, in each sucrose solution.
2 Plot a suitable graph and comment on your results.

Suggestions for further work

1 Investigate the germination of pollen grains in sucrose solutions with and without boric acid.
2 Investigate the effect of different concentrations of boric acid on germination. It is suggested that concentrations of boric acid between 0.001 and 0.01 per cent should be used.
3 Use an eyepiece graticule to measure the **rate** of growth of pollen tubes, measuring the length of the tubes every 30 minutes for 3 hours.
4 Make a preparation to show germinating pollen grains on the stigma. Place a stigma and about 1 cm length of the style from a mature flower on a microscope slide. Add a few drops of aqueous methylene blue solution (0.01 per cent). Apply a coverslip and press gently to squash the preparation. Leave for a few minutes, blot off excess stain, then examine. The pollen grains and tubes should stain darker blue than the cells of the stigma.

Changes in pH during the conversion of grass into silage

Introduction

The process of making silage is described in Chapter 4. In this practical, we investigate changes in pH which occur as fresh grass is converted into silage. Bacteria, which are naturally present in grass, convert sugars into organic acids, such as lactic acid, so the pH should fall.

WEAR EYE PROTECTION (Benedict's test)

Materials

- 1 dm^3 plastic lemonade bottle, or similar
- Freshly cut grass clippings. *Depending on where grass is harvested from (e.g. public places) it may be contaminated with animal faeces. Hygiene is therefore important.*
- Plastic fermentation lock and stopper to fit bottle
- Mortar and pestle
- Distilled water
- Benedict's reagent
- Test tubes and holders
- Universal indicator solution

Method

1 Grind up some of the grass clippings with a little distilled water, using a mortar and pestle.
2 Test some of the extract with Benedict's reagent and Universal indicator solution. Record the pH of the extract.
3 Pack the plastic bottle with grass clippings, pressing them down well to exclude as much air as possible.
4 Place some Universal indicator solution in the fermentation lock, then attach this to the bottle.
5 Leave the experiment at room temperature. Remove and test some of the contents with pH indicator solution every 4 or 5 days, for about 5 weeks. At the end of the experiment, test some of the contents for the presence of reducing sugars.

Results and discussion

1 Present your results in a table and plot a graph to show how the pH changed during the course of the experiment.
2 What gas was given off during the experiment?
3 Account for the changes which occurred in the grass during the conversion of grass into silage.

4 Farmers sometimes use additives, such as Ecosyl™ or Ecobale™, when making silage. Find out why these are added.

Identification of food constituents in milk

Introduction

The purpose of this practical is to identify and, where possible, quantify the food constituents of milk. The concentration of reducing sugars can be determined semi-quantitatively using Benedict's reagent and a range of colour standards. There are a number of possibilities for comparing the content of different types of milk, and of milk treated in different ways, such as pasteurised, sterilised and UHT.

Materials

- Samples of different types of milk
- Benedict's reagent
- Standard glucose solutions: 5.0, 2.0, 1.0, 0.5, 0.1 and 0.05 per cent
- Biuret reagent: dissolve 8 g of sodium hydroxide in 800 cm^3 of distilled water; add 45 g of potassium sodium tartrate and dissolve; then add 5 g of copper sulphate, dissolve, and add 5 g of potassium iodide; finally make up to 1.0 dm^3 with distilled water. Each reagent must be fully dissolved before adding the next. The solution should be kept in a dark bottle
- Sudan III in ethanolic solution
- Beaker to use as boiling water bath
- Pipettes or syringes
- Test tubes
- Microscope slides and coverslips
- Microscope

Method

1 To produce a range of colour standards, use a series of glucose solutions of known concentration. Add 0.3 cm^3 of each of these solutions to a series of appropriately labelled tests tubes, each containing 5.0 cm^3 of Benedict's reagent. These test tubes should then be placed in a boiling water bath for 8 minutes, then left to cool in air.

HARMFUL copper sulphate

IRRITANT sodium hydroxide

HIGHLY FLAMMABLE sudan (III) in ethanolic solution

WEAR EYE PROTECTION

2 Estimate the reducing sugar content of the milk samples, by pipetting 5.0 cm^3 of Benedict's reagent into a test tube and adding 0.3 cm^3 of the sample to be tested. Heat in a boiling water bath for 8 minutes, leave to cool, then compare the colour produced with the colour standards.

3 To test for proteins, place 2 cm^3 of the sample to be tested in a test tube and add an equal volume of biuret reagent. A purple-violet colour develops slowly, the intensity of which is proportional to the protein content.

4 To show the presence of fat, add a minute drop of Sudan III solution to a drop of milk on a microscope slide and apply a coverslip. Examine using a microscope; an emulsion of fat droplets should be visible.

Results and discussion

1 Prepare a table to record your observations using each test.

2 Compare the reducing sugar content, protein and fat content of each sample of milk.

Further work

1 The relative density of milk samples can be compared by determining the time taken for a drop of milk to fall through a solution of copper(II) sulphate. A layer of copper proteinate forms around the drop, which prevents the milk dispersing. Use a 100 cm^3 measuring cylinder, filled approximately 5 cm above the 100 cm^3 mark with 0.1 mol dm^{-3} copper(II) sulphate solution. Introduce one drop of milk, using a syringe fitted with a needle, just below the surface of the copper(II) sulphate solution and record the time taken for the drop to fall between the 100 cm^3 and 10 cm^3 marks. Repeat using different types of milk.

The resazurin test, methylene blue test and turbidity test

Introduction

The tests in this practical are to investigate the freshness of milk, by using methods which indicate the activity of bacteria, and to investigate the effectiveness of pasteurisation and sterilisation.

Resazurin is an indicator which shows metabolic activity of bacteria. The indicator is blue in the oxidised state but changes, when reduced, through pink to white. Although this test does not show the types of bacteria present, it can be used as a means of comparing the bacterial content of milk samples. Tubes containing milk which change colour to white, pink or white mottling have failed the test.

Methylene blue is a sensitive redox indicator which, like resazurin, shows bacterial activity in a milk sample. Methylene blue is decolourised when reduced, so recording the time taken for the blue colour to disappear gives an indication of bacterial activity in the milk sample.

The turbidity test is used to check for the efficiency of sterilisation. The procedure depends on changes in the properties of milk proteins after treatment at different temperatures and after the addition of ammonium sulphate. If, after addition of ammonium sulphate and filtration, the filtrate remains clear on boiling, the sterilisation procedure has been effective.

(i) The resazurin test

Materials

- Milk samples
- Resazurin tablets
- Distilled water
- Pipettes or syringes
- Sterile, screw-capped containers, such as universal bottles
- Water bath at 37 °C

IRRITANT
resazurin
tablets

WEAR EY
PROTECTIO

Method

1 Dissolve one resazurin tablet in 50 cm^3 of distilled water.

2 Add 1.0 cm^3 of this solution to 10.0 cm^3 of milk to be tested in a sterile container. Replace the lid, label the container and invert once to mix the contents.

3 Incubate in a water bath at 37 °C, filled so that the level of water is just over the level of milk in the container.

4 Set up a control tube containing 10.0 cm^3 of boiled milk plus 1.0 cm^3 resazurin solution.

5 Examine the samples after 10 minutes and note any colour changes.

6 Replace in the water bath and examine again after 1 hour.

7 Compare the colour of each sample with that of the control tube, which should remain blue.

(ii) The methylene blue test

Materials

- Milk samples
- 5.0 per cent acetaldehyde (ethanal) solution. Add a few drops of phenolphthalein indicator, followed by a dilute solution of sodium carbonate until the mixture just turns pink.
- 0.01 per cent methylene blue solution.
- Distilled water
- Pipettes or syringes
- Test tubes
- Aluminium foil
- Water bath at 40 °C

Method

1 Measure 5.0 cm^3 of pasteurised milk into a test tube, add 1.0 cm^3 of the acetaldehyde solution and 1.0 cm^3 of methylene blue. Mix the contents by shaking the tube gently, then cover the top of the tube with a small piece of aluminium foil.

2 Stand the tube in a water bath at 40 °C and note the time taken for the methylene blue to become decolourised. A blue ring may remain at the top of the sample.

3 Repeat this procedure with other samples of milk.

(iii) The turbidity test

Materials

- Milk samples: sterilised; pasteurised; pasteurised and boiled for 5 minutes
- Ammonium sulphate
- Electronic balance
- Conical flasks, 50 cm^3
- Measuring cylinder, 100 cm^3
- Filter funnels and filter paper
- Test tubes
- Beaker to use as a boiling water bath
- Pipettes or syringes
- Bench lamp

Method

WEAR EYE PROTECTION

1 Weigh 4 g of ammonium sulphate and transfer to a conical flask.

2 Add 20 cm^3 of the milk sample to be tested to the ammonium sulphate.

3 Shake the flask for at least 1 minute to dissolve the ammonium sulphate.

4 Leave the flask to stand for 5 minutes.

5 Filter the contents of the flask and transfer 5.0 cm^3 of the filtrate to a test tube.

6 Place the test tube in a boiling water bath and leave for 5 minutes.

7 Cool the tube in a beaker of cold water, then examine the contents by holding the tube in front of a bench lamp.

Results and discussion

1 Record all your results in a suitable table.

2 Compare the results of each test for the different samples of milk used and comment on their significance.

3 Find out about the possible health risks associated with untreated (raw) milk. What steps are taken to minimise these risks?

Investigating the structure and tensile strength of wool

Introduction

Wool consists of two main types of fibres, hairy fibres and true wool. Hairy fibres have a hollow core, referred to as the medulla, and are divided into long hair and shorter, coarser kemp. Hairs are the longest fibres present in the fleece. True wool has finer fibres, generally without an internal medulla. These different types of fibres are illustrated in Figure P.4.

The aim of this practical is to examine wool using a microscope, and to investigate the tensile properties of wool.

Materials

- Samples of wool, including a piece of yarn approximately 35 cm long

- Microscope fitted with eyepiece graticule
- Microscope slide and coverslip
- Cork cut in half lengthwise
- Retort clamp and stand
- Weights
- Ruler

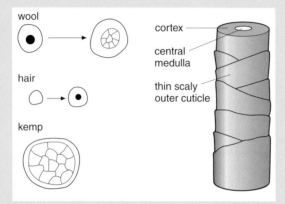

Figure P.4 *Different types of wool fibres. Top: showing how the medulla varies with the thickness of the true wool fibre. The middle diagram shows a fine fibre without a medulla and a coarse fibre with a non-latticed medulla. Kemp fibres have a wide, latticed medulla. Hairs are never as thick as kemps. The diagram on the right shows the main parts of a wool fibre.*

Method

1 Examine a few wool fibres using a microscope and measure their diameters. If present, identify the different types of fibre.
2 Clamp a loop of wool yarn securely, using a split cork and retort stand, as shown in Figure P.5.
3 Add weights progressively to the wool, recording the length after each weight has been added.
4 Remove the weights progressively and again measure the length of the wool each time.

Results and discussion

1 Prepare a table to summarise your findings about the types of wool fibres present in the sample.
2 Tabulate your data for the tensile strength experiment, including the extension of the wool as each weight is added.
3 Plot a graph to show the extension of the

wool yarn against the weight applied, for both loading and unloading. Does the wool return to its original length?
4 Comment on the shape of your graph.

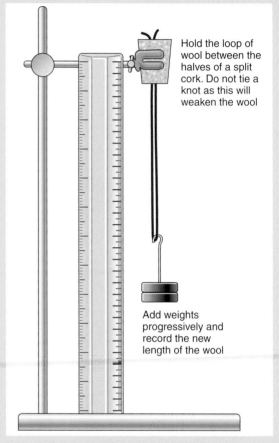

Hold the loop of wool between the halves of a split cork. Do not tie a knot as this will weaken the wool

Add weights progressively and record the new length of the wool

Figure P.5 *Method for investigating the tensile properties of a length of wool*

Suggestions for further work

1 If available, investigate the proportions of the different types of wool fibres present in samples of wool from different breeds of sheep.
2 Compare the tensile properties of wool with those of a synthetic fibre.
3 Investigate the effects of treatments such as boiling, or washing with detergents, on the tensile properties of wool.

Investigations into the structure of an egg
Introduction

The aim of this practical is to investigate the structure and composition of a hen's egg, including the identification of food constituents. Quantitative comparisons could be made between eggs produced in, for example, extensive and intensive systems.

Materials

- Egg
- Benedict's reagent
- Biuret reagent: dissolve 8 g of sodium hydroxide in 800 cm^3 of distilled water; add 45 g of sodium potassium tartrate and dissolve, then add 5 g of copper sulphate, dissolve, and add 5 g of potassium iodide; finally make up to 1.0 dm^3 with distilled water. Each reagent must be fully dissolved before adding the next. The solution should be kept in a dark bottle
- Sudan III in ethanolic solution
- Dilute hydrochloric acid
- Beaker to use as boiling water bath
- Pipettes or syringes
- Test tubes
- Microscope slides and coverslips
- Microscope
- Crystallising dish or other suitable container
- Forceps

HIGHLY FLAMMABLE Sudan (III) ethanolic solution

IRRITANT dilute hydrochloric acid

WEAR EYE PROTECTION

Method

1 Support the egg in a crystallising dish, using a crumpled paper towel. Crack the shell and carefully remove pieces with forceps, until a window has been made enabling you see the contents.
2 Make a drawing of your opened egg, labelling the **shell**, **shell membranes**, **air space**, **albumen**, **chalazae**, and **yolk** surrounded by **the vitelline membrane**.
3 Separate the albumen from the yolk and carry out tests to compare the content of protein, fats and reducing sugars. The methods for these tests are described in the practical: *Identification of food constituents in milk*.
4 Add a piece of shell to a small volume of dilute hydrochloric acid in a test tube.

Results and discussion

1 Record your results for the food tests in a suitable table, so that the contents of the albumen and yolk can be compared.

2 What happened when the shell was added to dilute hydrochloric acid? What does this indicate about the composition of the shell?
3 What other substances are present in the shell?
4 Prepare a flow chart to show the development of a hen's egg as it passes down the oviduct.

Suggestions for further work

1 Use a screw micrometer to measure accurately the thickness of the shells from a number of eggs. You could investigate whether or not there is a significant difference between the shell thicknesses of, for example, brown and white eggs or whether there is a significant difference in the thickness of the shell in the polar and equatorial regions of the egg.
2 Investigate the relationship between volume and mass of a sample of eggs. The volume of an egg may be found by using the displacement method with a measuring cylinder.
3 The shell of an egg is porous to allow gas exchange. Investigate the porosity of the shell by measuring changes in mass of a sample of eggs over a period of time.

Flower structure, bees and honey

Introduction

Many flowers have conspicuous markings on their petals, known as **nectar guides**, which insects recognise. These lead them towards the nectaries to obtain nectar. In this practical, we look at the structure of a flower with such nectar guides. Comparisons are made between worker bees, drones and queens, and honey is analysed to investigate its sugar content.

Materials

- Flowers to observe nectar guides. Flowers of the family Scrophulariaceae, such as foxgloves (*Digitalis purpurea*) or *Calceolaria* spp. are suitable
- Single-edged razor blades (*take care with these*)

PRACTICALS

WEAR EYE PROTECTION

- Bees
- Hand lens or stereo microscope
- Honey
- Distilled water
- Benedict's reagent
- Standard glucose solutions, 2.0, 1.0, 0.5, 0.1, 0.05, 0.02 and 0.01 per cent
- Diabur 5000 glucose test strips
- Test tubes
- Beaker to use as boiling water bath
- Syringes or graduated pipettes

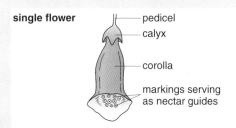

single flower
- pedicel
- calyx
- corolla
- markings serving as nectar guides

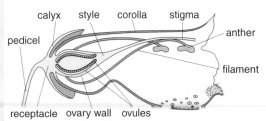

half flower
- calyx
- style
- corolla
- stigma
- pedicel
- anther
- filament
- receptacle
- ovary wall
- ovules

Figure P.6 Single flower, half flower and raceme of a foxglove.

Method

1 Examine the flower provided for nectar guides. Cut the flower in half lengthwise and make an annotated half flower drawing to show how it is adapted for insect pollination.
CAUTION: Handle foxgloves with care as they are poisonous.

2 Examine the bees using a hand lens or stereo microscope. Copy and complete Table P.1 to compare the three types.

Table P.1 *Summary table to show some differences between the three types of honey bee*

Feature	Worker	Drone	Queen
Overall length / mm			
Length of abdomen / mm			
Maximum width of abdomen / mm			
Pollen comb present on hind legs			
Mouthparts modified for moulding wax			

3 Investigate the total reducing sugar content of a sample of honey, and the glucose content. Concentrations of reducing sugars can be determined semi-quantitatively using Benedict's reagent and a range of colour standards. Quantitative estimations of glucose concentrations may be determined using suitable test strips, such as Diabur 5000.

4 To produce a range of colour standards, use a series of glucose solutions of known concentration. Add 0.3 cm^3 of each of these solutions to a series of appropriately labelled test tubes, each containing 5.0 cm^3 of Benedict's reagent. These test tubes should then be placed in a boiling water bath for 8 minutes, then left to cool in air.

5 To estimate the concentration of reducing sugars in the honey samples, it will be necessary to make a suitable dilution of the honey, using distilled water. A 1 per cent solution (w/v) should be tried first, and adjusted if necessary. Pipette 5.0 cm^3

of Benedict's reagent into a test tube and add 0.3 cm^3 of the solution to be tested. Heat in a boiling water bath for 8 minutes, leave to cool, then compare the colour produced with the colour standards.

6 If using Diabur test strips, a strip should be dipped into the solution to be tested, removed, and the colours produced compared with the colour chart after 2 minutes. This method is specific for glucose, and will give quantitative results. Remember to multiply your result by the dilution factor to obtain the original concentration of glucose.

Results and discussion

1 Collate your drawing of the flower, summary table to show the morphology of the three types of honey bees, and results of your investigation into the sugar content of honey.
2 Find out how you could measure the sucrose content of a honey sample.
3 Write a brief concluding summary of the interdependence of flowering plants and honey bees.

Suggestions for further work

1 Compare the glucose contents of different samples of honey.
2 Describe a method by which the different sugars present in honey could be separated and identified using paper chromatography.

Examination questions

Chapter 1

1 Fertilisers are applied to the soil to improve the
nutrient supply to crops and thus increase yields.
Organic fertilisers, consisting of plant and animal
remains, have been used for hundreds of years. A
widely used organic fertiliser is farmyard manure
(FYM).
  (a) (i) Give two advantages of using farmyard
          manure rather than inorganic fertilisers.
                                    (2 marks)
      (ii) Give two disadvantages of using farmyard
           manure.   (2 marks)
  (b) Green manuring is another method of adding
      organic matter to the soil.
      Describe how green manuring is carried out
      and suggest how it is of benefit to the farmer.
                                (3 marks)
                           (Total 7 marks)

2 An investigation was carried out into the effects of
carbon dioxide concentration on yield. Tomato
plants were cultivated in glasshouses, where it was
possible to control the concentration of carbon
dioxide in the atmosphere.
The carbon dioxide concentrations ranged from
50 to 1200 parts per million (ppm) by volume. The
yield of tomatoes was measured in kg per m². The
temperature and light intensity conditions were
constant for all concentrations of carbon dioxide.
The results are shown in the graph below.

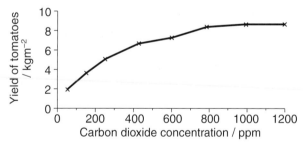

  (a) (i) Describe the effects of increasing the
          carbon dioxide concentration on the yield
          of tomatoes.   (2 marks)
      (ii) The normal concentration of carbon
          dioxide in the atmosphere is approximately
          300 ppm. From the graph, determine the
          yield of tomatoes when the concentration
          in the glasshouse was 300 ppm.   (1 mark)

      (iii) Calculate the percentage change in yield
          that would be expected if the tomatoes
          were grown in an atmosphere where the
          carbon dioxide concentration was
          increased to 800 ppm compared with the
          yield at 300 ppm. Show your working.
                                  (2 marks)
  (b) Explain why carbon dioxide concentration
      affects the yield of tomatoes   (2 marks)
  (c) A further experiment was carried out to
      investigate the effect of temperature on yield.
      In this experiment the carbon dioxide
      concentration was 300 ppm and the light
      intensity remained constant.
The results are shown in the graph below.

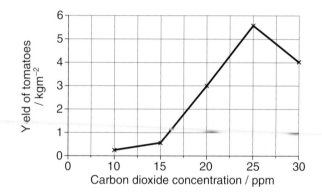

Compare the yield of tomatoes at 15 °C with that at
25 °C and suggest an explanation for the difference
in yield.   (3 marks)
  (d) Name one factor other than light, temperature
      and carbon dioxide concentration, which could
      affect the yield of tomatoes grown in a
      glasshouse.   (1 mark)
                         (Total 11 marks)

Chapter 2

1 The interaction between weeds and crop plants was
investigated using field barley and the weed, white
persicaria (Polygonum lapathifolium). Seeds of both
were sown together in pots and seedlings of both
plants emerged on the same day.
The table below shows the numbers of barley plants
and persicaria plants in each of four experiments.
Large numbers of replicates of each experiment
were made.

Experiment	Number of barley plants	Number of persicaria plants
A	1	4
B	1	128
C	4	1
D	128	1

At weekly intervals from the time of emergence the mean dry mass per plant of barley surrounded by 4 or 128 persicaria plants and the mean dry mass of persicaria surrounded by 4 or 128 barley plants were determined.

The results are shown in the graph below.

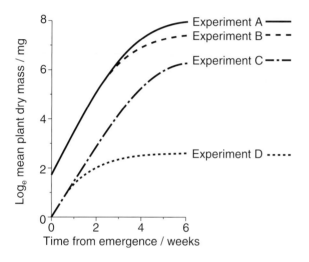

(a) Explain why the mean dry mass of the plants was determined in these experiments.

(3 marks)

(b) (i) Compare the results of experiment A (1 barley and 4 persicaria) with the results of experiment B (1 barley and 128 persicaria). (2 marks)

(ii) How do the results from experiments C and D, where the weed is surrounded by the barley, compare with the results of experiments A and B? (3 marks)

(iii) Suggest two conclusions which can be drawn from the results of these experiments. (2 marks)

(c) State two characteristics of annual weeds which contribute to their success.

(2 marks)

(Total 12 marks)

2 Aphids feed on crops, such as beans, and may also spread viral diseases.

An investigation was carried out into the relationship between the environmental temperature and the number of aphids in active flight. Experiments were carried out at temperatures from 10 °C to 26 °C. At each temperature twenty readings were used to calculate a mean percentage of aphids in flight.

The results were shown in the graph below.

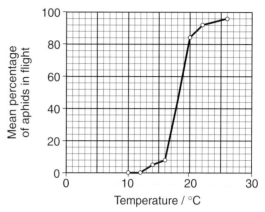

(a) (i) At what temperature would 50% of the aphids be in flight? (1 mark)

(ii) Twenty readings were also taken at 15 °C and 25 °C and the means and standard deviation calculated. The results are shown in the table below.

Temperature / °C	Mean percentage of aphids in flight	Standard deviation
15	8	4.05
25	96	2.01

Suggest which of the two readings is the more reliable, giving a reason in support of your answer.

(2 marks)

(b) (i) Describe the effect of temperature on the flight activity of the aphids. (2 marks)

(ii) Suggest an explanation for this effect.

(3 marks)

(c) (i) Suggest two advantages of flight to aphids.

(2 marks)

(ii) The bean plants on which the aphids feed are annuals. Explain how the aphids survive through the winter months. (2 marks)

(Total 12 marks)

3 A fruit grower wanted to protect a pear crop from fungal disease, so a programme of spraying the trees with a fungicide was devised.

EXAMINATION QUESTIONS

The following sequence was recommended.

Stage of growth	Treatment
Bud burst	Spray
Green cluster	Spray
White bud	Spray
Blossom	No spray
Petal fall	Spray
Fruitlet (mid-June)	Spray
Fruitlet (early July)	Spray

(a) (i) Suggest why the fruit grower was recommended to spray the trees at the different stages rather than at specific dates. (2 marks)

(ii) Suggest two reasons why the trees were not sprayed at blossom time. (2 marks)

(b) The fruit grower has a choice as to whether a systemic or a protectant fungicide is used.

(i) State two ways in which a systemic fungicide differs from a protectant fungicide. (2 marks)

(ii) Give two disadvantages of using protectant fungicides (2 marks)

(c) There are 55 pear trees in the orchard. The recommended treatment for each tree is 5 to 10 dm³ of fungicide spray. Calculate the minimum amount of spray needed to give protection for a whole season.
Show your working. (2 marks)

(d) Fungi can quickly become resistant to systemic fungicides.
Suggest two precautions that the fruit grower can take to prevent development of resistance to the fungicide. (2 marks)

(Total 12 marks)

Chapter 3

1 The practice of artificially propagating plants by taking soft-tip cuttings is widely used in horticulture.

(a) Give one example of a plant which is commonly propagated commercially by soft tip cuttings. (1 mark)

(b) Describe how this technique is carried out (3 marks)

(c) State two advantages to commercial growers of this method of propagation (2 marks)

(Total 6 marks)

2 The diagram below shows some stages in the micropropagation of a crop plant.

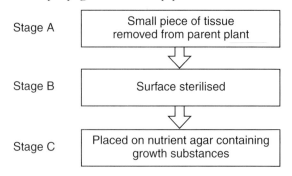

(a) Name a tissue suitable for use at stage A (1 mark)

(b) (i) What can be used to surface sterilise the tissue at stage B? (1 mark)

(ii) State two precautions that would need to be taken to avoid contamination after surface sterilisation. (2 marks)

(c) Name one plant growth substance which would be included in the nutrient agar used at stage C. (1 mark)

(Total 5 marks)

3 Give an account of pollen formation and explain how controlled pollination can be used in plant breeding.

(Total 10 marks)

Chapter 4

1 The table below gives descriptions of some of the characteristic features of mammalian herbivores related to their nutrition. Complete the table by stating the name of the structure to which each description refers.

Description	Structure
Allows continual growth of teeth	
Permits efficient grinding and crushing of plant food	
Allows manipulation of food in the mouth during chewing	
Is the site of cellulose digestion by microorganisms	

(Total 4 marks)

2 Resazurin is a dye used to test the effectiveness of different methods of milk preservation.
In an investigation, samples of raw fresh milk, pasteurised milk and sterilised milk were tested. 10 cm³ of each milk sample was combined with 1 cm³ of resazurin solution. The tubes were stoppered, shaken and incubated at 37 °C for several hours and then examined

(*a*)   State why the tubes were stoppered.   (1 mark)

(*b*)   Suggest *two* precautions which would be taken to ensure the investigation is reliable.
(2 marks)

(*c*)   (i)   State the colour changes you would expect to see in the sample of raw milk.
(1 mark)

(ii)   State which of the samples listed above would be likely to show the longest storage time and give a reason for your choice.   (2 marks)
(Total 6 marks)

3 Read through the following account of wool production in sheep, then write on the dotted line the most appropriate word or words to complete the account.
'Wool and hair fibres are made of a protein called keratin. In domestic sheep, there are three main types of fibres, produced from .... in the skin. Brittle white fibres, known as ...., are usually .... in length and have .... Hairy heterotypes are very variable in length and coarseness, the finest being indistinguishable from wool. Fine wool fibres are produced from .... and do not have a ....
(Total 6 marks)

4 The diagram below shows how *in vitro* fertilisation of eggs from cattle can be carried out.

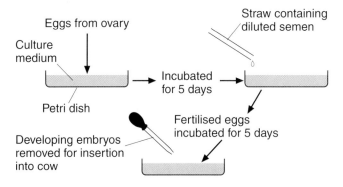

(*a*)   (i)   Give one way in which a cow can be treated to stimulate the release of several eggs at the same time.   (1 mark)

(ii)   State why the eggs are kept in a culture medium for 5 days before semen is added.
(1 mark)

(*b*)   (i)   Give *two* ways by which the developing embryos may be inserted into the uterus of a cow.   (2 marks)

(ii)   Name the stage of the cow's oestrous cycle at which this should be carried out.
(1 mark)

(iii)   Some of the embryos are kept to be used later. State how they could be stored in a healthy condition until they are used.
(1 mark)
(Total 6 marks)

5 *Sward height* (the height of the vegetation) is a useful practical indicator of the availability of *herbage* (grass and other plants) to grazing animals. Sward surface height can be measured by placing a ruler vertically so that the lower edge just touches the ground. The height of the tallest leaf at that point is then recorded.
The graph below shows the influence of sward surface height on milk production in kg day⁻¹ from spring-calving cows.

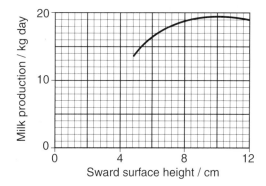

(*a*)   (i)   Describe the relationship between milk production and the sward surface height.
(2 marks)

(ii)   Using the information on the graph, suggest a suitable minimum sward height needed for grazing lactating dairy cows.
(1 mark)

(*b*)   The table on page 132 (top) shows the minimum sward heights needed for some other classes of grazing animals.

(i)   Comment on and suggest an explanation for the difference between the minimum sward height needed for ewes and lambs and that for flushing ewes.   (3 marks)

Class of grazing animal	Minimum sward height / cm
Ewes and lambs	4 to 5
Flushing ewes	6
Heifers and dry cows	6 to 8
Sucklers with calves	7 to 9
Beef cattle	7 to 9

(ii) Suggest a reason for the difference in minimum sward heights for sheep and cattle. (1 mark)

(c) Suggest and explain two factors which could affect the height of the sward on land used for grazing. (4 marks)

(Total 11 marks)

6 Beef is produced from bulls and from steers (castrated bulls).

An investigation was carried out to compare the production of beef from bulls and steers fed in different ways. One set of bulls and steers was fed on a silage (fermented grass) based diet and a second set was fed on a cereal based diet.

The table below gives the results from the investigation.

Feature	Silage based diet		Cereal based diet	
	Bulls	Steers	Bulls	Steers
Gain per day / kg	1.54	1.27	1.89	1.40
Carcass mass / kg	272	254	280	253
Age at slaughter / days	359	377	321	253
Total cost / £	245	255	290	286
Gross profit per animal / £	285	240	256	207
Gross profit per hectare / £	633	533	731	591

Adapted from Crabtree, et. al. (1988) Fast Finishing of Suckler Bulls, British Cattle Breeders Club Winter Conference, 1988

(a) Suggest why the growth rate for the bulls was higher than for the steers. (2 marks)

(b) Calculate the percentage difference in carcass mass of bulls fed on the silage based diet compared to steers fed on the same diet. Show your working. (2 marks)

(c) Suggest two reasons why the animals fed on the cereal based diet had a faster growth rate than those fed on the silage based diet. (2 marks)

(d) (i) Compare the relative merits of keeping bulls on a cereal based diet as opposed to a silage based diet. (3 marks)

   (ii) Suggest why the gross profit per hectare was higher for animals fed on a cereal based diet. (2 marks)

(e) Suggest one advantage and one disadvantage of using steers rather than bulls for beef production. (2 marks)

(Total 13 marks)

7 Give an account of the conversion of grass into silage.

(Total 10 marks)

Chapter 5

1 The diagram below shows the reproductive system of a female chicken and the parts involved in egg production.

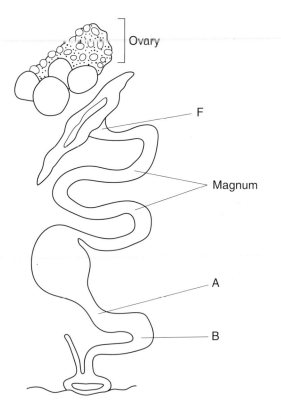

(a) State the function of the magnum (1 mark)

(b) Describe what happens to the egg as it passes through A and then through B. (3 marks)

(c) On the diagram show by means of an arrow and the letter F the region in which the ovum is fertilised. (1 mark)

(Total 5 marks)

2 A poultry farmer kept a record of the mean number of eggs laid per day over a period of 12 months by a flock of free-range chickens.
The results are shown in the graph below.

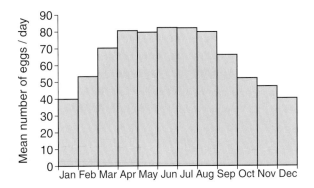

(a) (i) Explain what is meant by the term free range. (2 marks)

(ii) Suggest two disadvantages of free range egg production. (2 marks)

(b) Comment on the variation in the mean number of eggs produced throughout the period of 12 months. (3 marks)

(c) Suggest how a high rate egg production could be maintained on a poultry farm. (1 mark)

(d) Suggest and explain two factors which might affect the size of an egg laid by an individual hen. (4 marks)

(Total 12 marks)

Chapter 6

1 A reliable method for the production of triploid rainbow trout has been developed. It involves subjecting newly fertilised eggs to a mild heat shock. This suppresses the meiotic division of the egg nucleus and results in the eggs remaining diploid. A diploid egg combining with a haploid sperm produces a triploid zygote.
A comparison was made of the growth of normal diploid female trout reared in the same tank with triploid female trout. Fifty fish of each kind were caught and weighed at four-weekly intervals during the second and third year of their development. The results are shown in the graph below.

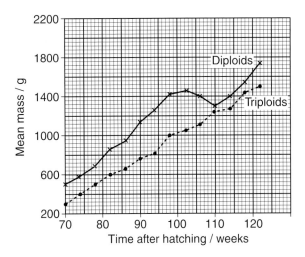

(a) (i) Compare the growth rates of the two groups from week 70 to week 100 and suggest an explanation for any differences you observe. (4 marks)

(ii) Suggest explanations for the differences in the growth rates of the two groups during the period week 102 to week 110. (3 marks)

(b) Suggest an advantage to the fish farmer of raising triploid rainbow trout. (2 marks)

(c) Suggest two ways in which fish farming is of benefit to the customer. (2 marks)

(Total 11 marks)

2 Give an account of the management of the life cycle of fish in fish farming.

(Total 10 marks)

Chapter 7

1 Give an account of the ways in which the activities of honey bees are of economic importance to humans.

(Total 10 marks)

Mark schemes

EDEXCEL Foundation, London Examinations accepts no responsibility whatsoever for the accuracy or method of working in the answers given.

In the marks schemes the following symbols are used:
; indicates separate marking points
/ indicates alternative marking points
eq. means correct equivalent points are accepted

Chapter 1

1 (a) (i) Any *two* from wide variety of nutrients ; improves soil structure ; improves aeration ; improves water retention ; increases cation exchange ; lower cost / cheaper ; (2 marks)

(ii) Any *two* from bulky / heavy / difficult / unpleasant to handle ; may contain weed seeds ; slow release of nutrients / takes time to break down ; variable nutrients / not much phosphorus / nutrient content not known / ref. to C/N ratio ; not local / may have to be transported ; smelly ; (2 marks)

(b) crop grown in the autumn / on bare land / ref. to remains of crop / eq. ; ploughed in / left on ground and next crop directly drilled in ; nutrients ; nutrients released slowly next / following spring ; reference to 'catch crop' / example of crop ; prevents erosion / leaching avoided ; (3 marks)

(Total 7 marks)

2 (a) (i) rapid increase initially / at low levels of carbon dioxide / up to 430 ppm ; then levels off / eq. ; not so much increase in yield above 700 / 800 ppm ; (2 marks)

(ii) 5.5 / 5.6 kg m^{-2} (1 mark)

(iii) $\frac{2.7}{5.5} \times 100$ / eq. ; 47.3% / eq. (2 marks)

(b) carbon dioxide needed / raw material for photosynthesis / eq. ; increase in carbon dioxide increases rate of photosynthesis /eq ; (2 marks)

(c) large / eq. increase in yield ; yield × 11 greater / increases from 0.5 to 5.5 kg / eq. ; increase in temperature increases rate of enzyme reactions ; light independent stages of photosynthesis enzyme controlled / eq. (3 marks)

(d) water / mineral ions / named ion / presence of pollinating insects / disease / variety ; (1 mark)
(Total 11 marks)

Chapter 2

1 (a) mean is more reliable / accurate ; dry mass a more accurate measure of growth / eq. ; plants may contain variable amounts of water ; dry mass indicates increases due to synthesis / eq. ; (need to take mean as single plants may vary / not be typical / individual differences eliminated ; (3 marks)

(b) (i) rate of growth of barley similar / eq. for first two / two and a half weeks ; less growth in B / B levels off sooner / converse for A ; (2 marks)

(ii) similar pattern of increase shown by C / all four ; dry mass of C + D lower / less than A + B or less / lower rate of growth of weed / converse for barley ; much less / eq. growth of weed in D / levels off early / ref. to time / eq. ; growth in D affected earlier than in B where same density ; growth in C parallel to A / B but lower / ref. to figures ; (3 marks)

(iii) interspecific competition / eq. is shown in both ; the higher the density of competing plant / e.g. / the greater the effect / eq. ; the growth of barley is affected less than the growth of the weed / barley competes better / Persicaria more vulnerable to competition ; (2 marks)

(c) Any two from large numbers of seeds / eq. ; intermittent germination / eq. ; shorter life cycle / rapid reproduction / germinate rapidly / no dormancy ; plasticity / eq. ; (2 marks)
(Total 12 marks)

2 (a) (i) 18.5 °C (allow between 18.0 and 18.5) ; (1 mark)

(ii) the mean at 25 °C / 96% is the more reliable ; standard deviation is greater at 15 °C / less at 25 °C ; any given reading at this temperature likely to vary more from mean / eq. / accept converse ; (2 marks)

(b) (i) slow increase up to 16 °C / no aphids in flight at 10 °C ; very rapid / eq. increase in numbers flying from 16 °C to 20 °C / most become active between 16 °C and 20 °C ; nearly all aphids / 96% in flight at 26 °C ; *General point*: increase in temperature increases % age of aphids in flight / increases flight activity ; (2 marks)

(ii) aphids ectothermic / poikilothermic / body temperature varies with external temperature ; metabolic rate / enzyme activity depends on temperature ; need (lots of) energy for flight ; so rate of respiration needs to be fast enough / eq. ; only possible above minimum temperature / ref. to actual figure from graph ; (3 marks)

(c) (i) dispersal / migration / colonise / new areas / avoid competition / eq. ; move to other bean plants for food / eq. ; finding mates ; move to overwintering plants / trees ; (2 marks)

(ii) migrate to tree / shrub / winter host ; ref. to mating / fertilised eggs produced ; eggs remain until spring / overwintering reference ; (2 marks)
(Total 12 marks)

Chapter 3

1 (a) chrysanthemum / geranium / fuchsia / eq. (1 mark)

(b) use this year's growth / juvenile / non-woody / eq. ; stem cut just below a node / eq. ; put into compost / suitable medium / eq. ; high humidity / misting / description / eq. ; (3 marks)

(c) fast / saves time / large numbers possible / eq. ; genetically identical to parent / clone / eq. ; can be done straight into pots for selling / all year round / eq. ; (2 marks)
(Total 6 marks)

2 (a) meristematic tissue / cambium / apical bud ; (1 mark)

(b) (i) sodium hypochlorite / bleach / eq. (1 mark)

(ii) use of sterile forceps / instruments for handling / ref. to sterile apparatus / equipment ; rinsed in sterile water ; carried out in sterile laminar flow cabinet / eq. ; put into sterile medium / sterile agar / eq ; (2 marks)

(c) auxin / cytokinin / kinetin ; (1 mark)
(Total 5 marks)

3 *Formation of pollen*: microspore mother cells ; in the pollen sacs (of the anthers) ; divide by meiosis ; haploid cells develop into pollen grains ; ref. to two-layered wall ; exine / intine ; exine pitted / sculptured if insect-pollinated / smooth / eq. for wind pollination ; (haploid) nucleus of pollen grain undergoes mitosis ; to give generative nucleus and pollen tube nucleus ; generative nucleus undergoes a further division to produce two male / gamete nuclei ;
Controlled pollination: anthers / stamens removed from one plant / ref. to emasculation ; self pollination before pollen ripe / eq. ; prevents self-pollination / eq. ; pollen from desired / selected / eq. plant ; transferred by paintbrush / eq. ; to stigma of emasculated flower ; has to be done when stigma / carpel mature ; flower covered / eq. ; to prevent other pollen getting in ; may carry out reciprocal crosses ; seeds formed collected ; reference to choice of characteristics / eq. ;
(Total 10 marks)

Chapter 4

1

Description	Structure
Allows continual growth of teeth	open roots of teeth ;
Permits efficient grinding and crushing of plant food	ridges on cheek teeth / eq. / circular jaw movements ;
Allows manipulation of food in the mouth during chewing	diastema / description ;
Is the site of cellulose digestion by microorganisms	rumen / caecum ;

(Total 4 marks)

2 (a) to prevent air / oxygen entering / prevent oxidation ; (1 mark)

(b) ref. to sterilisation of glassware / eq. ; repetition of the experiment / use of larger number of samples ; thermostatic control of temperature / ref. to keeping the temperature constant / use of water bath ; stated time for incubation ; complete exclusion of air from

MARK SCHEMES

apparatus ; stirring / shaking at regular / eq. intervals   **(2 marks)**

(c) (i) change of colour from blue / purple to pink / white / colourless ;   **(1 mark)**

(ii) sterilised milk ; greater number of bacteria destroyed than by pasteurisation / all bacteria / spores killed / eq. ;   **(2 marks)**

(Total 6 marks)

3 follicles ; kemp ; short / eq. ; pointed tips / tapering roots / medulla / eq. ; secondary follicles ; medulla / hollow centre / eq.

(Total 6 marks)

4 (a) (i) with follicle-stimulating hormone / FSH / gonadotrophins / PMSG ;   **(1 mark)**

(ii) to make sure they were alive / healthy / viable eq. ;   **(1 mark)**

(b) (i) *through* / *via* the vagina / cervix ; surgically (through wall of abdomen and uterus ;   **(2 marks)**

(ii) just after ovulation / during oestrus / luteal phase / eq.   **(1 mark)**

(iii) by keeping them in liquid nitrogen / at −196 °C.   **(1 mark)**

(Total 6 marks)

5 (a) (i) increase in milk production as sward height increases ; reaches plateau / levels off above 8 / 9 / 10 cm / above a certain height increase in sward height does not increase yield / ref. to slight decrease ;   **(2 marks)**

(ii) 8 / 9 / 10 cm   **(1 mark)**

(b) (i) higher / eq. for flushing ewes / lower for ewes and lambs / or ref. to figures ; flushing ewes need better quality / more food for a few weeks before ovulation / increases follicle production / eq. ; increases the incidence of twins ;   **(3 marks)**

(ii) cattle bigger need more food / sheep smaller / eq. / cattle cannot graze as low / eq.   **(1 mark)**

(c) density of grazing / type of animal grazing / eq. ; more animals the shorter the grass / rotating use of fields / trampling effect ; mineral nutrition / nutrients in soil / amount of fertiliser used ; growth of grass needs adequate nitrogen / eq. supply ; situation / location / ref. to mountain / moorland eq. ; growth affected

by climatic conditions ; nature of grass / species / mixture / eq . ; some varieties grow to different heights / eq. ;   **(4 marks)**

(Total 11 marks)

6 (a) more testosterone / androgens / growth hormones present / converse ; more production of bone ; more production of muscle / fat ;   **(2 marks)**

(b) $\frac{272-254}{254} \times 100$ $\frac{18}{254} \times 100$; 7.1 %   **(2 marks)**

(c) contains more energy / converse for silage ; contains more protein / converse for silage ; easier to digest / converse for silage ; more nutritious = 1 mark (if neither of the first two given)   **(2 marks)**

(d) (i) cereal fed bulls have highest growth rate / gain more weight / day ; cereal fed bulls have highest carcass mass ; cereal fed bulls have greatest profit per hectare ; cereal fed bulls reach slaughter age earlier ; silage fed bulls have the highest profit per animal ; silage fed bulls have the lowest production cost ;   **(3 marks)**

(ii) less land needed per animal / more animals per hectare ; less land needed for growing cereals than for growing grass for silage ; food richer in energy / protein ; slaughtered earlier ;   **(2 marks)**

(e) *Advantage*: steers easier to manage / less aggressive ; *Disadvantage*: steers grow more slowly / lower carcass mass / less profit per hectare ;   **(2 marks)**

(Total 13 marks)

7 grass cut (into short pieces) ; cut in afternoon as sugar content higher ; easier to compact into silo ; chopping breaks open cells / makes cell contents available for bacterial action ; use of forage harvester to cut and chop the crop ; blown into trailers ; may be grass cut into swathes with mower ; then allowed to wilt ; picked up and chopped ; wilting concentrates the sugars ; cut crop placed in silo in layers ; compacted / packed tightly ; sealed with plastic sheet / eq. ; conditions anaerobic / eq. ; bacteria convert sugars ; to lactic acid ; lowers pH / pH goes down to 4–4.5 ; preventing further microbial deterioration ;

(Total 10 marks)

Chapter 5

1 (*a*)   secretes / adds / eq. albumen / eq. ;

(1 mark)

(*b*)   fluid added ; ovum swells ; calcium salts / pigment / cuticle added to cell ; shell shell membrane formed / added ;   (3 marks)

(*c*)   labelling line accepted from just above infundibulum to just above magnum

(1 mark)

(Total 5 marks)

2 (*a*)   (i)   chickens kept in the open / eq. ; feeding not controlled / eq. ;   (2 marks)

(ii)   less control over diseases ; vulnerable to predators ; seasonal laying ; fewer eggs ; birds affected by weather ; eggs may be dirty if laid outdoors ; more land needed / economical point / eq. ;   (2 marks)

(*b*)   lowest numbers produced in winter months / eq. ; highest numbers April to August / high numbers maintained / stays around 80 / ref. to peak ; increase January to April / decrease August to December ; production from April to August double that of production in December to January ; corresponds with better weather / more light / availability of food / daylength ;

(3 marks)

(*c*)   maintain good feed / maintain high light intensity / long daylength ;   (1 mark)

(*d*)   *Factor*: age of hen
Explanation: older hens lay larger eggs ; breed of hen / eq. bantams lay small eggs / eq. ;
Factor: sequence in clutch ;
Explanation: eggs produced at the beginning of clutch larger ; (protein content of) feed / diet ; more protein, larger eggs ; climatic conditions ; high temperatures reduce egg size
(4 marks)

(Total 12 marks)

Chapter 6

1 (*a*)   (i)   both show steady increase in growth / both growing at same rate ; triploids always lighter than diploids / converse ; credit for attempt to work out rate ; triploids could be genetically smaller / diploids genetically bigger / size a genetic factor ; triploids not so good at competing for food / diploids better at getting food ;

(4 marks)

(ii)   triploids continue growth increase, diploids lose mass ; (diploids lose mass) because lose eggs ; reproduction ref. ; comment on how much mass loss / figures quoted ; triploids sterile / infertile / no eggs produced ;   (3 marks)

(*b*)   triploids continuous increase in mass / growth / all nutrients used for growth / diploids not gaining mass for 14 weeks ; comment on better quality / better if not spawning ; can predict when triploids ready to sell ; demand for large fish could be seasonal / smaller fish more saleable ;   (2 marks)

(*c*)   continuous supply of trout / salmon / not dependent on seasons / all year round ; size not variable / can be harvested at optimum / required / desired size ; quality guaranteed / disease-free / eq. ; not depleting ocean fish stocks / less hazardous ; mass production gives lower price ;   (2 marks)

(Total 11 marks)

2   many fish will not spawn in captivity ; may only spawn at certain times of year / brought into season by increase / eq. in daylength ; possible to induce spawning by hormones / ref. to hypophysation ; to ensure production of fish all year round ; hormones extracted from pituitary glands of fish used ; used on mature fish ; fertilisation under controlled conditions ; ref. to stripping to obtain gametes / eq. ; details of stripping / ref. to massaging abdomen / eq. ; fertilisation details / use of dry container / eq. ; embryos transferred to incubator ; ref. to shading from light ; ref. to temperature management at any stage of development ; left undisturbed for 10-15 days ; when eye-spots visible eggs shocked by moving to another container / eq. ; to kill damaged / unviable eggs ; fry transferred to small holding tanks ; ref. to frequent feeding / encourage fry to feed ; ref. to growing on to maturity ; ref. to broodstock for breeding / not all fish allowed to breed ; most fish harvested at suitable size for market ; ref. to development of all female fish / giving females male hormones ; reference to development of triploids / eq.

(Total 10 marks)

MARK SCHEMES

Chapter 7

1 bees as pollinators of crops / eq. credit example of crop / apples / field beans / fruit / eq. ; reference to beehives in orchards / eq. ; used as pollinators inside glasshouses ; visit flowers for nectar / pollen ; ref. to honey guides / bee brushing against anthers ; pollen on body carried to next flower visited ; cross pollination if compatible / eq. ; honey source of human food / sweetener / eq. ; honey formation described / ref. to regurgitation by worker bees / removal of water from nectar / medicinal properties ; disinfectant / counteracts inflammation ; wax used for cell building in hive / comb / eq. ; useful to humans in polishes and waxes ; other products credited e.g. coatings / cosmetics / candles / adhesives – allow any two for 2 marks ; ; royal jelly as a health food / eq. ; source of vitamin E / antioxidant ; pollen used as a medicine / help for anaemia / eq. ; healing properties of propolis / fights infections / eq. ;

(Total 10 marks)

Index

INDEX